Eight-Legged Marvels

Beauty and Design
in the
World of Spiders

Chad Arment

COACHWHIP PUBLICATIONS

LANDISVILLE, PENNSYLVANIA

Eight-Legged Marvels, by Chad Arment

Copyright © 2008 Chad Arment

Coachwhip Publications

CoachwhipBooks.com

ISBN-10 1-930585-40-3

ISBN-13 978-1-930585-40-9

Images
Front Cover: *Argiope* orbweaver (© Anthony Berenyi)
Back Cover, top right, clockwise: orbweaver with stabilimenta (© Alexander Shalamov), jumping spider (© Clarence Seet), crab spider (© Christian Musat), orbweaver (© Ismael Montero), wolf spider (© Iva Villi), Ecuadorian spider, unknown sp. (© Ra'id Khalil), running crab spider (© Eric Coia), red-headed mouse spider (© Gilles Paire).
Title Page: huntsman spider (© Ryan Pike)

CONTENTS

4

It isn't difficult to determine that the outline of a Spider sprawling across the Nazca desert in Peru is the result of intelligence in design and action rather than chance activity of the natural elements. We know that this geoglyph was created, though we don't know for certain who created it or even why. But what about a living spider? The debate over design and purpose in nature, particularly whether philosophical naturalism is the only legitimate scientific perspective on the natural world, continues unabated.

Nazca Lines, Peru
© Jarno Gonzalez Zarraonandia

Giant Lichen Orbweaver,
Araneus bicentenarius
© Lew Scharpf

Cupiennius getazi, Panama
© Dr. Arthur Anker, Smithsonian Tropical Research Institute, Panama City, Panama

INTRODUCTION

A tarantula's careful progress along the rainforest floor, a garden spider's spinning-work, a trapdoor spider's swift pounce-and-capture from a hidden burrow—spiders captivate and intrigue us. Some of us enjoy watching them, while others jump at the sight of their eight-legged march.

However they strike us, spiders are a choice introduction to biology as we explore diversity in form, function, and behavior within the natural world. Spiders also give us opportunity to consider the philosophies behind worldviews held by different people within science. Someone who believes strictly in materialistic naturalism and someone who believes in a personal and involved Creator see the same colors, patterns, structures, and behavior in spiders, yet often interpret these details differently when discussing purpose, beauty, or worth.

It is common for philosophical naturalists to limit the concept of natural design to organized functionality. Dr. Francisco Ayala asserts that design in living organisms is simply the result of natural selection creating and preserving novel genetic adaptations out of chance variations. Some critics suggest this is out-of-order—that design precedes selection (good designs are preserved, etc.). There are also arguments that an undirected evolutionary process has limited capabilities for creating novel genetic adaptations, and that biological systems showing irreducible or specified complexity evidence actual, rather than apparent, design.

Is design illusory in nature? Or is it purposeful, giving us insight into the creativity and orderliness of God? Is nature's beauty only in the eye of the beholder? Or should we extend our aesthetic sensibilities beyond superficial loveliness and toward a glimpse of an original creation God pronounced, *very good*? Spiders give us ample opportunity to consider such questions. What follows is a basic biological overview of spiders and their world, with the hope that it offers a starting point for discussions on beauty and design in nature.

Wolf Spider
© Cathy Keifer

Tarantula, *Pterinochilus* sp.
© Tomasz Slowinski

Beauty and Design in the World of Spiders

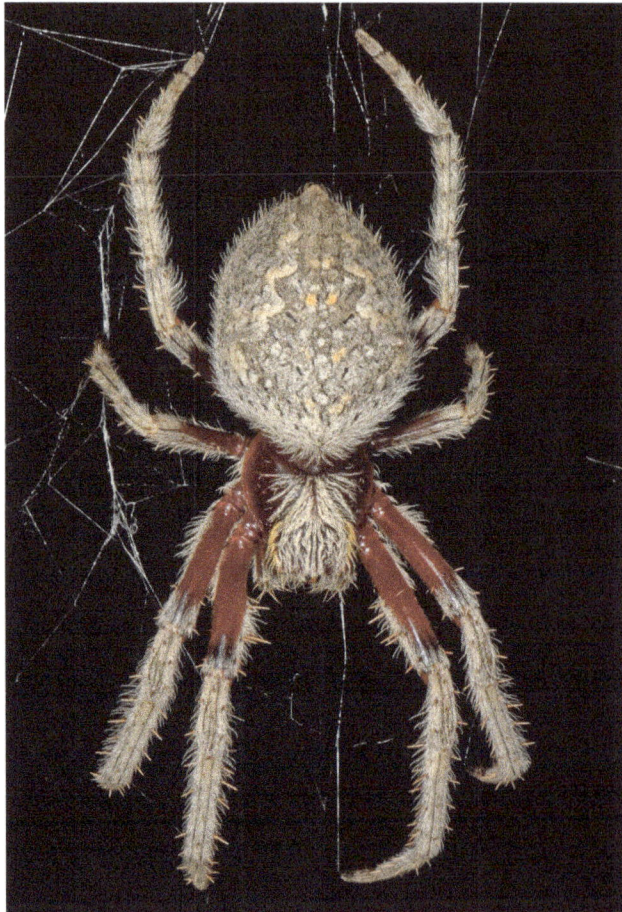

What is beauty? The painter draws his "line of beauty," and points to the graceful attitude of the Grecian statue,—the politician speaks of utility and pleasing associations. We shall allow metaphysicians to settle the dispute as best they may. All allow that there is such a thing as beauty, and that closely allied to it, though in some respects differing from it, there is also something else, to which we give the name of grandeur or sublimity. There is, moreover, a corresponding emotion in the mind, which finds its appropriate gratification in contemplating those objects in which these qualities are exhibited. A very large portion of our mental enjoyment arises from the pleasurable feelings that are in this manner produced. And our great and gracious Father, in storing the universe around us with such a rich and varied supply of the objects to which we have referred, and in bestowing on us emotions by which we are led to take pleasure in their contemplation, has clearly manifested the benevolence of His designs, and calls on us to rise with adoring delight from Nature up to Nature's God.

— Rev. James Brodie,
The Rational Creation, 1855

Orbweaver, *Argiope* sp.
© Howard Cheek

Orbweaver, *Eriophora* sp.
© Stuart Elflett

WHAT ARE SPIDERS?

Spiders are invertebrates, meaning their body is principally supported by an exoskeleton. The spider's chitinous exterior is tough, whether hardened (sclerotized) or flexible, providing stability as it moves. (Mammals and other vertebrates are supported from the inside by a bony or cartilaginous endoskeleton.) Spiders are also ectothermic, so external sources or ambient temperatures regulate their body heat.

Being arachnids, spiders share certain characteristics with invertebrates like scorpions, harvestmen, solifugids, ticks, and mites. All arachnids have four pairs of true legs, a pair of palps or pedipalps (which in scorpions have claws), and two body segments. (This segmentation may be difficult to discern externally in groups like mites.) One characteristic sets the order Araneae apart from other living arachnids—specialized glands producing silk that is spun by the spider's spinnerets.

About 40,000 species of spiders have been described, but estimates suggest there may be up to five times as many species alive today. Certain characteristics distinguish three broad groupings of living spiders. These are commonly termed the Mesothelae, the Mygalomorphae, and the Araneomorphae.

Theridion with Egg Sac
© Kevin Pfeiffer

Mesothelid Segmentation
Image: Comstock (1912)

Mesothelids are well-known from the fossil record, but only a few—the Liphistiidae—remain today. Liphistiids are burrowing trapdoor spiders from southeast Asia. While tarantula-like in posture and form, liphistiids are distinguished by a segmented abdomen and having eight spinnerets located centrally underneath the abdomen.

Mygalomorphs include the true tarantulas and trapdoor-spiders. Their fangs thrust up and down, and they have two pairs of book-lungs. Their primary spinnerets are found at the tip of the abdomen.

Araneomorphs are the "typical" spiders, but vary widely in shape and behavior. Approximately 90% of all known spiders are araneomorphs. Their fangs meet crossways, and they have no more than one pair of book-lungs.

A Small Brazilian Spider
© Achilles Moreaux

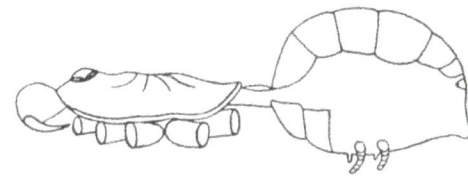

Beauty and Design in the World of Spiders

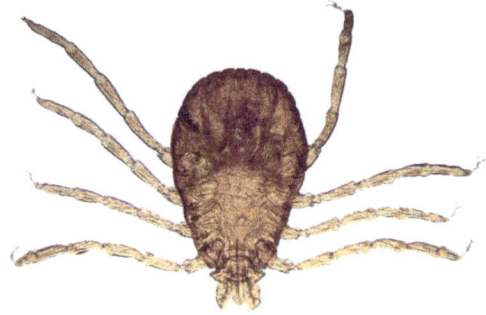

Other Arachnids

Top Right, Clockwise:
Dog Tick (© Carolina K. Smith, M.D.);
Pseudoscorpion Hitching Ride on a Beetle
(© Ryszard Laskowski); Scorpion (© Nico
Smit); Velvet Mite (© Michael Pettigrew);
Harvestman (© Mihail Orlov)

Opposite:
Tailless Whipscorpion (© Ra'id Khalil);
Giant Vinegaroon (© Robert Barber);
Camel Spider or Wind Scorpion (© Casey
Bishop)

Beauty and Design in the World of Spiders

SPIDER ANATOMY

While insects have three-segment bodies (head, thorax, and abdomen), spiders and other arachnids have two segments: the cephalothorax (or prosoma) and the abdomen (or opisthosoma). In the cephalothorax, the spider's head is fused to its thorax. A sturdy carapace above and sternum below protect this segment, sheltering the spider's brain and much of the digestive system. The cephalic groove in the middle of the carapace is called the fovea. This is used as an attachment point for internal muscles.

Spiders cannot swallow whole prey. They crush their food with their mouthparts (chelicerae), while liquefying it with digestive enzymes, or suck fluids from prey after piercing it with their fangs. The chelicerae are a pair of small muscular appendages directly in front of the mouth. In most species, the chelicerae contain venom glands, fangs, and cuticular teeth.

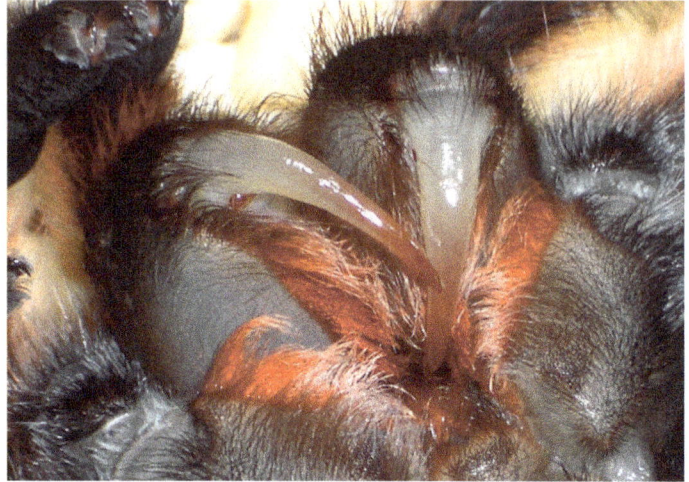

Tarantula Fangs
© Warren Rosenberg

There are two fang configurations in spiders. Diaxial fangs, seen in araneomorph spiders, move crossways, pincing against each other. Mygalomorph and mesothelid spiders have paraxial fangs that thrust up and down, parallel to each other.

Spiders have up to four pairs of eyes, arranged variously over the carapace. A few species, like recluse spiders and some uloborids, have only two or three pairs. Rarer are species with only one pair of eyes or, as with some cave spiders, no eyes at all. Most are simple eyes, but jumping spiders, wolf spiders, and certain others have unusually complex eyes. The simplest eyes detect motion and light intensity, while others identify patterns in polarized light, recognize a small range of colors, or amplify light. The complex eyes found in some species identify potential mates, competitors, or prey by shape and behavior.

Jumping Spider, Brazil
© Euclides Neri

Four pairs of legs and a pair of palps emerge from under the cephalothorax. Legs are used in such activities as locomotion, prey capture, web-spinning, and courtship. The palps are short leg-like appendages found in front of the true legs. They manipulate food and sense vibrations. In mature male spiders, the palps are configured for reproductive purposes. Trichobothria are specialized hairs found on certain parts of the legs and palps. These thin hairs each sit within a cup-like socket, sensing even the lightest motion of air or ground vibration. Sound-detecting and pressure-gauging slit sense organs, as well as chemical "tasting" receptors, are also found on the legs.

The cephalothorax connects to the abdomen by a pedicel. This "waist" is more or less noticeable depending upon the species, though easily identified when viewed from underneath the

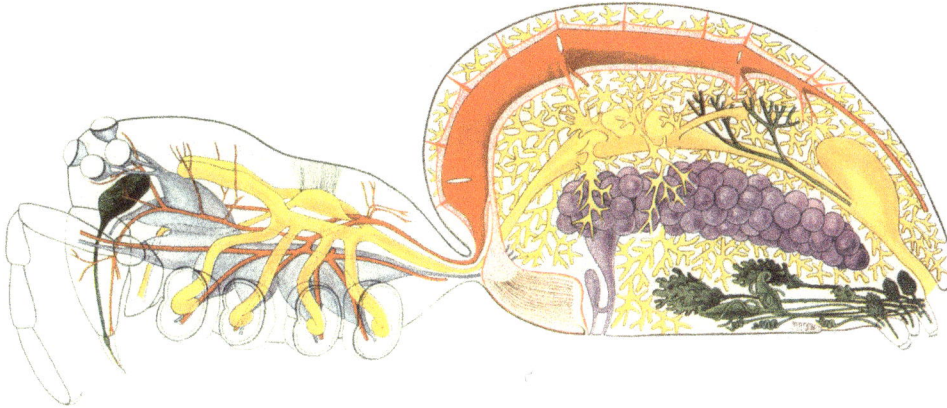

Internal Anatomy of a Spider

Body Wall and Respiratory Organs (Black), Vascular System (Red), Alimentary Canal (Yellow), Nermous System (Blue), Reproductive Organs (Purple), Silk Glands, Poison Glands, Malphigian Vessels (Green), Venom Gland (Green, in Cephalothorax). Image: Comstock (1912).

spider. The anterior aorta, nerves, and a portion of the stomach pass through it. In all but a few species, females are larger than males, with proportionally larger abdomens. Internally, abdomens house components of the circulatory, respiratory, reproductive, digestive, excretory, and silk-spinning systems.

While spiders rely on their exoskeleton for overall support, they do have an internal support structure made up of connective tissue. This structure is called the endosternite and acts as point of attachment for different muscle groups. It also controls the fluid pressure of the spider's blood, extending and flexing leg joints that lack muscular control. When a spider dies, the fluid pressure to extend those joints is lost, so the legs curl up.

The circulatory system centers on the tubular heart. When the heart contracts, the blood (or hemolymph) pushes forward and backward into the anterior and posterior aortas. Valves keep the blood from re-entering the heart. Each aorta branches out into arteries running throughout the body. When blood leaves the closed arterial system, it enters the open venous system. Rather than closed veins like we have, the spider's system works via low-pressure mechanisms, where the pumping action of the heart creates pathways in the free-flowing blood. This ensures that blood circulates past all tissues, including respiratory organs, on its way back to the heart. After contraction, the heart swells back to normal size, creating suction that pulls venous blood in through slits called ostia. As the heart

Bird-Dropping Mimic
© Farhan Bokhari

Beauty and Design in the World of Spiders

contracts, these slits close and the blood continues to circulate.

Spiders vary in their respiratory systems. Pholcid spiders use a single pair of book-lungs, mygalomorphs and mesothelids have two pairs of book-lungs, while a pair each of book-lungs and branching tracheae are found in most other spiders. Some very small spiders require only a pair of tiny sieve tracheae. The spider's blood delivers oxygen and carbon dioxide back and forth between the respiratory organs and tissues. The book-lungs collect air, allowing gaseous diffusion to occur with the blood through a thin membrane. Tracheae deliver oxygen deeper within the spider's tissues, but still require circulating respiratory pigments in the blood to complete respiration. Openings (spiracles) to the internal tracheae and book-lungs can be seen on the underside of a spider's abdomen.

Araneomorph (*Palpimanus gibbulus*)
© Jorge Almeida

Spinnerets are found at the tip of the abdomen (or under the abdomen, in mesothelids). These connect to silk glands, which often fill up a great portion of the abdomen. Besides the silk glands (and glands which add adhesive liquids to silk), other secretory glands are known to be present, though not all have been identified. Pheromones for mate attraction, chemical prey attractants, and chemicals that induce social tolerance have been found in various spiders.

Spiders grow by molting their skin. The process starts as enzymatic secretions separate the old skin from the underlying epidermal cells. After about a week, new underlying skin begins to form. When this is finished, the old skin splits, and the spider extracts itself. In some cases, a missing leg, spinneret, or other appendage regenerates. At maturity, males are ready to mate, having acquired the necessary reproductive equipment during this last molt. Only a few spiders ever molt after maturity, as they rarely survive the season.

Size in spiders ranges from one extreme to another. Certain species of the Symphytognathidae are usually considered the smallest spiders. The full-grown male of the Columbian species *Patu digua* has been measured with a body size of only 0.37 mm. But, the female *Anapistula caecula* from the Ivory Coast is as small as 0.46 mm. The male should be smaller, though none has yet been found—leading to speculation that the species is parthenogenic. The heaviest spider known is *Theraphosa blondi*, the goliath bird-eating tarantula. A full-grown adult can weigh over 120 grams and have a legspan of about twelve inches. A few other tarantulas may have slightly longer legspans, though are not as heavy. Araneomorph spiders can have large legspans—*Heteropoda maxima* from Laos reaches 300 mm in legspan—but do not have a tarantula's girth. Scientists used to believe the largest spider ever was the fossil species *Megarachne servinei,* with a twenty-inch legspan. A closer examination of the fossil, however, showed that it wasn't a spider at all, but an unusual eurypterid, or sea scorpion.

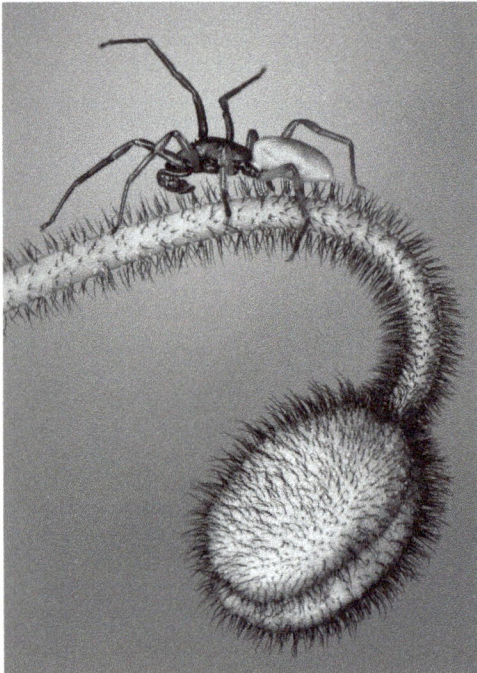

Tiny Spider on Poppy Stamen
© Cathy Keifer

Appearance

A spider's colors and patterns affects its ability to survive in the wild. Many camouflaged spiders blend in so well with bark, sand, or other natural elements that they won't be noticed unless they move. They hide from both predators and their own prey. In some spiders, camouflage isn't readily apparent to human eyes, as it may be based on light wavelengths we cannot see.

Other spiders show warning, or aposematic, coloration—bright colors designed to alert predators that the spider is dangerous or distasteful. Some biologists suggest that the bright colors of *Argiope* orb-weavers warn birds from flying into their expansive webbing.

Not all bright colors are warning signals, however. With certain orb-weaving spiders, bright colors attract potential prey. In jumping spiders, colorful patches are used in courtship displays, though those spiders don't have the ability to detect a full color range.

Some spiders are chameleon-like, able to change hues instantaneously. Sometimes this is used defensively, as when the linyphiid spider *Floronia bucculenta* darkens when disturbed. Certain crab spiders use their color-changing ability to adapt to flower coloration as they wait in ambush for pollinating insects.

Baboon Spider
© Nico Smit

Gasteracantha arcuata, Borneo
© Dr. Arthur Anker, Smithsonian Tropical Research Institute, Panama City, Panama

Female Black Widow
© Steve Potwin

Ant Mimic, Thailand
© Chartchai Meesangnin

A Red Ant Mimic
© James Benet

Mimicry in color, form, and behavior is known in several spider families. Some mimic objects like wood knots, bird droppings, or twigs. Others mimic species like beetles, ants, and pseudoscorpions. There are several possible reasons for a species to mimic another animal. Spiders fall into multiple recognized categories for mimicry of other organisms.

Batesian mimicry: the spider mimics an animal that is distasteful or dangerous. Some beetles have tough shells or secrete repugnatorial chemicals, so predators may avoid a spider that mimics a beetle.

Müllerian mimicry: both the spider and the animal it mimics are distasteful or dangerous, reinforcing a predator's disinterest.

Peckhamian or *aggressive mimicry*: the spider mimics a species that it preys on. Some ant-mimicking spiders feed on ants themselves. Mimicry allows the spiders to get close enough to pounce without warning.

Wasmannian mimicry: this is a specialized form of mimicry, pertaining to spiders and other species that mimic ants in order to live safely within an ant colony.

Schooling mimicry: the spider mimics a species that is found in groups (like ants), allowing it to evade predators by blending in with the crowd.

REPRODUCTION

The reproductive openings (the female's epigynum and the male's gonopore) are situated centrally underneath each spider's abdomen. A mature male spider is also equipped with distinctive bulbous organs at the end of the palps. Each of the bulbs is tipped with a long tube called the embolus. In some species, these bulbs are so large the spider appears to wear boxing gloves. The tiny males of one cobweb spider, *Tidarren sisyphoides*, have massive bulbs that slow them down. Before their last molt, males twist off one palp, allowing them to survive longer in their search for a mate.

Male spiders must prepare themselves for mating by filling up these specially configured palps with

House Spider with Eggsacs
© Loren Lewis

sperm. First, the male spider spins threads or a small web on which he directly releases a droplet of sperm from his gonopore. The male draws up the sperm into the palpal organ, possibly through capillary action, where it remains until he locates a suitable mate.

After courting a willing female, the male inserts the embolus into the female's epigynum. In mygalomorph and haplogyne araneomorph spiders, this embolus is a simple tapering "sting"-like tube. In entelegyne araneomorph spiders, however, the embolus has a complex coil arrangement. This coil matches the female's internal sperm storage organ (spermathecae) just as a key fits its lock. Each species has its particular matching configuration. This specialization is one trait by which such species can be distinguished. Arachnologists examine the reproductive organs in order to accurately identify entelegyne

Pair of Bolas Spiders, with Eggsac
Cladomelea akermani
© John Roff, Hilton College, South Africa

Palps on a Male Spider
© Sabrina dei Nobili

Beauty and Design in the World of Spiders

Female Wolf Spider
Carrying Spiderlings
© Samuel Maissonnier

Spiderlings,
Araneus diadematus
© Wilmar Huisman

spiders, because so many other morphological characters are highly variable. With some spiders, the embolus breaks off during mating and plugs the reproductive tube, so that the male ensures his paternity.

After mating, the female begins to develop eggs. Depending upon the species, the eggs may be laid within weeks or months. The eggs are fertilized as they are laid. Most spiders wrap the egg mass together in a silk cocoon. The cocoon protects the eggs against the elements, parasites, and predators. The cocoon may be camouflaged for hiding, protected within further webbing, or carried around by the mother. If eggs are laid early in the season, they will hatch within a few weeks. Later clutches may overwinter before hatching. Some female spiders lay more than one clutch after a single mating.

While there are exceptions, most male spiders die within weeks of mating. Many females die the same year they lay their eggs, though sometimes living long enough to care for the spiderlings. Those spiders that hatch, grow, reproduce, and die in a single year are univoltine species. Spiders with overlapping generations, when adults or spiderlings overwinter as necessary, are multivoltine species.

FEEDING AND VENOM

Spiders live in many habitats, and capture a wide range of prey. Insects like moths, grasshoppers, and bees are particularly relished, but prey sometimes has a sting or mandibles that must be carefully avoided during capture. Some spiders attack ants and beetles in the face of poisonous or irritating chemicals, or even earthworms, snails, and slugs that use slime for protection. Aggressive spiders are known to tackle dangerous prey: dragonflies, wasps, and even other spiders. The largest spiders stalk and feed on lizards, frogs, fish, and baby birds. The strongest orb webs trap small birds and bats.

Cross Spider (*Araneus diadematus*) Feeding
© Gerard Maas

In almost all cases, the prey must be live and moving to trigger a feeding attack, but a few species (wolf spiders and certain small gnaphosids, like *Herpyllus*) will scavenge on dead prey.

Spider venom immobilizes prey, but also serves to defend the spider from other predators. These venoms are complex blends of peptides, proteins, polyamines, enzymes, and other components. Spider toxins are generally neurotoxic (attacking the nervous system) or cytotoxic (destroying cells and tissue), while a few have shown myotoxic (damaging muscles) or haemotoxic (attacking the blood) effects. Spider venom is not used to digest prey, however. Spiders must regurgitate digestive enzymes onto captured prey. As the meal liquefies, the spider's stomach creates suction like a miniature vacuum cleaner.

The toxicity of a particular venom component can be prey-specific. Within the black widow's venom, there are insect-, crustacean-, and vertebrate-specific latrotoxins. This allows the black widow to capture and feed on a wider range of prey than spiders that have single-target venoms. Specialization within venoms can provide evidence of the history for a spider group's recent lineage. All *Loxosceles* and *Sicarius* (recluses and kin) spider venoms contain a particular large protein, Sphingomyelinase D, which plays a significant role in the tissue-destroying nature of the venom. This protein is not found in any other animal. Though there is variance within the protein chain between species, particularly between those found in different regions of the world, the ever-present protein suggests a common origin of *Loxosceles* and

Red-Headed Mouse Spider (*Missulena occatoria*)
© Richard Hall

Sicarius. Of course, critical reasoning skills must always be applied to investigate whether a given trait found in different organisms truly reflects a common lineage.

Only a very small percentage of spiders has the ability to harm humans. Reports of dangerous spider bites often exaggerate the risk. Other medical conditions have been misdiagnosed as spider bites, particularly for suspected brown recluse bites. Necrosis of tissue is one possible symptom of a recluse bite, but can also be symptomatic of other problems. Too many times the recluse is implicated in regions where populations do not naturally exist. All spiders should be treated with respect, but there is little reason for outright fear.

Top Left, Clockwise:
Green Lynx Spider Capturing Wasp
© Rose Hayes

Orbweaver Catches a Dragonfly
© Kevin Karge

Orbweaver Wrapping Bee
© Louis M. Barbee

Top Left, Clockwise:
Scolopendrid Centipede Eating Spider, Malaysia
© Ang Sek Chuan

Green Huntsman Spider with Spider Prey
© Jorge Almeida

Wasp Carrying Dead Spider, Costa Rica
© Aliza Schlabach

Praying Mantis Feeding on Orb-Weaving Spider
© Alexander Chelmodeev

Beauty and Design in the World of Spiders

SILK

Spiders rely on silk for many of their activities: spinning webs, producing draglines and safety lines, traveling over bridge lines or ballooning through the air, laying prey detection and capture strands, wrapping and storing prey, protecting eggs with silk cocoons, creating shelters (including underground and underwater retreats), attracting prey or mates with chemically treated silk, creating sperm webs (for male spiders) and forming stabilimenta in orb-webs.

Spinnerets

Spinneret-silk gland complexes have three main parts. Inside the abdomen, a production sac secretes a liquid crystalline mixture of small proteins called fibroins. Attached to the sac are looped spinning ducts, where the proteins are aligned, folded, and stretched. When needed, the fibroin mix is pulled out of the ducts by the external spinnerets, rapidly hardening into silk. Spinnerets are clusters of microscopic spigots, usually seen under the tip of the abdomen. With a few exceptions (cribellar and epiandrous), these spigots are muscular, mobile, and fully innervated, with valves controlling the thickness of the silk. Spinnerets may have one, two, or many spigots, each capable of its own independent movement. Spinnerets weave silk fibers together as they are pulled from the spigots.

Spinnerets of an Orb-Weaving Spider (*Gasteracantha*)
© Dennis Kunkel Microscopy, Inc.

The fibroin proteins consist mainly of the amino acids glycine, alanine, and, in some silks, serine. The physical properties of silk are determined by the proportion of amino acids within a protein group, the proportion of proteins added to a strand, the rate of speed at which it is pulled from the spinnerets, its thickness, the amount of stretching it undergoes, the physical configuration of the spigots, and which glands contribute to it. In orb-weavers, for example, ampullate silk spigots are round, producing sturdy round silk threads. In recluse spiders, these spigots are flattened, producing wide, but extremely thin, ribbons of elastic silk.

While spiders often use specialized glands for particular purposes, some spiders are generalists and use whichever primary glands they have to complete the task at hand, even combining silk from different glands. There are many glands known to produce silk or glue, but most spiders have only a selection of them. One burrowing mygalomorph spider, *Antrodiateus unicolor*, has the simplest known structure, with only a single type of silk gland for females to use in lining their retreat and spinning an egg-sac.

The "typical" glands include:

Ampullate glands: The major ampullate glands are responsible for the spider's dragline that is attached to any surface the spider walks over. In some species, mature females add a chemical to the dragline that points potential mates in their direction. The dragline becomes a safety line when the spider drops from a height. Both major and minor ampullate silk are used to construct the framing lines and radii for prey capture webs. Ampullate silk is also used by female wolf spiders to attach their egg sac to their spinnerets.

Major ampullate silk is often touted for its physical qualities (strength and elasticity) when compared to artificial products. Minor ampullate silk is not as strong as major ampullate silk, and is very inelastic. When stretched out, it remains that way.

Polynesians, Native Americans, and other peoples have used major ampullate silk from the larger species of *Nephila* orb-weaving spiders as fishing lines and nets. In terms of tensile strength, major ampullate silk from the orb-weaver *Araneus* is not quite as strong as steel of the same length and diameter, and only a third as strong as a comparable fiber of Kevlar®. This spider silk is, however, far more flexible and extensible. These balance with strength to produce strands that are much tougher than steel or Kevlar® in their ability to absorb energy from an impact. (Being much lighter, a strict comparison of ampullate silk with a thread of steel of the same weight will show the spider silk to be much stronger.) Any comparison of spider silk to manufactured products, however, must keep in mind that spider silks are rarely meant to last longer than a few hours or days. Silk from the major ampullate glands, while initially strong, does not have long-term durability under constant stress.

In many species, major ampullate silk has an additional characteristic: supercontraction. When wet, the threads shrink to less than half their original length while swelling twice their original diameter. On a molecular level, the threads enter a more fluid state and reconfigure as hydrogen bonds are broken. This does not usually affect a finished orb-web, as the framework serves to hold each strand in place, but supercontraction plays a vital role during the formation of ampullate silk within the liquid environment of the silk glands. As the supercontracted protein chains pass through the spinning ducts, they are stretched out by application of the spinneret's friction brakes, realigning the chains. This creates a stronger silk. By varying the pressure of the valves, the spider modifies the strength and flexibility of the thread to fit the need.

Pyriform glands: these create small patches of short sticky silk (attachment disks) that bond other silk strands together (as when creating webs), or affix silk threads to another surface (as when attaching a safety line to drop from a height). Pyriform silk is bipartite—composed of a separate biochemical core and adhesive skin, rather than being one long strand of proteins.

Cucumber Green Spider
(*Araniella cucurbitina*)
© Marka Kosmal

Beauty and Design in the World of Spiders

(TL) Huntsman Spider Guarding Egg Sac on Bamboo, China
© Cedric Bassett

Argiope Orb-Weaver's Eggsac
© Patrick Lamont

Multi-Species "Giant Spiderweb" Covering Trees at Lake Tawakoni State Park, Texas, in 2007.
CC: Joe Lapp

Typical Orb-Web
© Roman Krochuk

Aciniform glands: these create wide swathes of tough silk for wrapping around prey. These glands also fashion the stabilimenta found in orb-webs.

Aggregate glands: these do not produce silk. Instead, they release liquid drops that, in orb-weaving spiders, are added to a pair of flagelliform silk threads to create sticky capture silk. Stickiness is directly proportional to the amount of glycoprotein granules present in the liquid. The aggregate gland liquid coats and maintains capture threads, both as a bactericide and by attracting moisture from the surrounding air. In certain sheet-web spiders, like the bowl-and-doily spiders, aggregate gland liquid is added between layers of densely packed mesh-webs. When the liquid evaporates, the webbing cements together.

Flagelliform glands: these produce the core spiral capture threads for orb-weaving spiders. Capture threads are highly elastic silk capable of stretching to twice their length, yet still return to their original shape. Elasticity is needed because the energy of a large insect hitting the orb-web must dissipate as the strands stretch or the web will break, allowing the insect to escape. These strands are not so resilient, though, as to bounce prey out of the web. (In three-dimensional space-webs, energy is dispersed by insects breaking multiple threads as they fly deeper into the webbing.) Capture threads need the aggregate gland liquid not only for its stickiness, but because it keeps the strands moist and stretchable. Uncoated capture threads have little elasticity and may fracture under stress. Aggregate liquid drops also act as reservoirs for spooled loops of threading as capture silks contract with moisture. Thus, when an insect hits the capture threads and pushes against them, the threads initially extend rapidly, but slow down with exponentially greater force as the threads stretch further. A corresponding mechanism is found at the molecular level, as models suggest the strands' protein chains are interconnecting spring-like arrangements.

Cribellar glands: these are found only in spiders with a specialized spinning organ called the cribellum. As liquid silk is secreted from the ducts, they pass through this hard plating, emerging from hundreds or thousands of tiny fixed spigots as finely spun microfibrils. These are spun together or coat an additional silk thread, then fluffed out by stiff serrated hairs on the spider's calamistrum, or "comb-foot." These cylindrical microfibrils capture prey both by snagging on hairs and by being "sticky." The inherent stickiness of cribellar threads is mostly due to electrostatic or van der Waals forces between the microfibrils and surfaces. Some microfibrils have nodes that take in atmospheric moisture and provide additional capillary or hygroscopic adhesion.

Pseudoflagelliform glands: in some cribellate spiders, such as the deinopids, these produce two to four axial fibers which support cribellar microfibrils.

Paracribellar glands: these produce silk fibers that are included by some species in their cribellar strands, linking the axial threads together. These appear to be modified aciniform glands.

Purseweb Spider Tube
Image: Comstock (1912)

Hummingbird Trapped in *Argiope* Web
© Bill Beatty

Beauty and Design in the World of Spiders

Orb-Weaver Wrapping Its Prey
© Marek Kosmal

Baby Orbweavers Migrating on Silk Strand
CC: Anissa Housley

Grasshopper, Wrapped and Stored
© Hway Kiong Lim

Lobed glands: these are found only in theridiid spiders with reduced aciniform glands. They, too, spin prey-wrapping silk.

Cylindrical or Tubuliform glands: these spin silk threads used to wrap eggs. Tubiliform silk is stronger than minor ampullate silk, but not as strong as major ampullate silk. Other silks (aciniform and ampullate) are usually incorporated into the egg-wrapping. In some species, like *Argiope* orb-weavers, the egg-sac silk coverings are paper-like. When spiderlings begin to emerge, they must regurgitate digestive enzymes on the silk. This dissolves other silks, allowing them to push the tubiliform silk aside and squeeze out of the egg-sac.

Epiandrous glands: these are found in male mygalomorphs and certain araneomorph spiders, near their gonopore. They are used only in spinning a sperm-web.

Cheliceral glands: technically not silk glands, these are found in the spitting spiders. They are located in the cephalothorax, and expel sticky glycoproteinous strands at prey and enemies.

As arachnologists study a wider variety of spiders, they encounter glands that do not conform to what are stereotyped as "typical" silk glands. This is particularly notable in non-araneomorph spiders. In the mygalomorph spiders of the diplurid genus *Euagrus*, their dry sticky webs are produced by a single gland type with only one kind of spigot. Electrostatic charge creates the stickiness in these strands.

Spiderling, Releasing Thread For Ballooning
© Steven Stone

Dispersal using Silk

Silk plays an important role in the dispersal of many species. In the simplest cases, spiders create their own walking-paths. Spiderlings of many species release a silk thread into the air that floats along until it sticks to another surface. The spiderling then walks the bridging line. They often do this several times in their search for a suitable territory.

Small araneomorph spiders (and a few mygalomorphs, like purse-web spiderlings) utilize silk for long-distance air travel, called ballooning. It is not uncommon for spiderlings to balloon *en masse*, sometimes creating long swathes of silk that fall from the sky. To initiate flight, the spider typically climbs to an elevated spot, sometimes spinning a silk line or platform from which to launch. It stretches high on its legs as if on tiptoes. Pointing its abdomen up, its spinnerets pull out a short slender length of ampullate silk. This gossamer silk catches on air currents, pulling out a further length. As the thread rapidly draws out, the moving air's drag on the silk and the spider's body creates enough force to lift the spider into the air.

Once in the air, the spider has little control over where it goes. How the spider positions itself in the air, however, may affect how quickly it lands. If it lands in an unsuitable spot, the spider can try again and drift elsewhere.

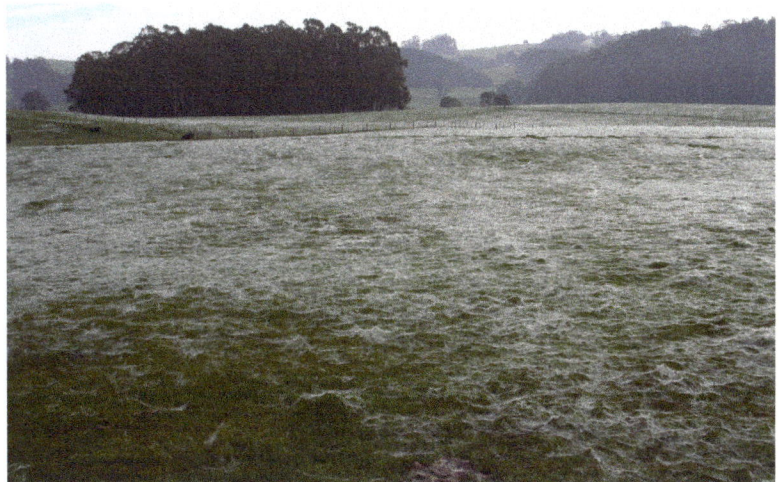

Field Covered in Gossamer
CC: Stephen Barnett

Hairy Mygalomorphs
Theraphosidae

These are what are commonly called tarantulas, though that term has been used to characterize any number of large hairy spiders. Some are actually tiny species, such as the Arizonan *Aphonopelma paloma*, with adults commonly under an inch in length. Other mygalomorphs are the largest known spiders. One record-holder was a Venezuelan *Theraphosa blondi* with a legspan of about 11 inches. Hairy mygalomorphs are found in both the Old and New World. Habitat preferences range from humid tropical regions to deserts and grasslands.

Some of the best-known tarantulas are *Brachypelma* species from Mexico and Central America. These are popular in the pet trade due to their size, colors, longevity, and often docile nature. Bright colors in this genus actually work as camouflage, blending in with vegetation and leaf litter. In the wild, these spiders dig permanent tubular burrows in the soil that contain separate chambers for feeding and molting. These tunnels are often several feet in length. Females that are ready for courtship will place strands of silk at the entrance of their burrow so that passing males can detect their presence and reproductive status. Males usually mature within 5 to 8 years, females within 10 years. Males actively search for a mate during what will usually be their sole adult year, while females may live another ten years. Rarely, a female lives more than thirty years.

The male's courtship is primarily vibratory. When a male *Brachypelma* detects a female through physical and chemical cues in her silk, he makes his first moves. He drums his legs against the silk on the female's burrow entrance, shakes his body, and beats his palps on the ground. As she approaches close enough for him to touch, he starts to box her with his palps. In many tarantulas, the female

Brachypelma sp.
© Juri Bizgajmer

Poecilotheria ornata
© Maigi Photography

Tarantula
© Tomasz Slowinski

displays threateningly, requiring the male to grip her fangs while he continues. If she is receptive, she arches her abdomen, allowing him to insert his palps from underneath. As they separate, the female may withdraw into her burrow, or she may attack the male. If she departs peacefully, the male sometimes employs deception to keep other males from visiting her, covering her burrow entrance with his own silk.

A few tarantula families are arboreal, stalking the many small creatures that inhabit the tropical treetops. They may spin silken retreats or find suitable space within a clump of vegetation or underneath tree bark. They remain still for hours, waiting patiently for prey to come into range, then strike quickly. Some of the larger arboreal tarantulas tackle small vertebrates, earning the nickname "bird-eating spiders."

While some species inflict painful bites, that isn't the only defense these spiders have. Most New World species rub or flick hairs on potential predators. These urticating hairs, usually found on the abdomen, irritate the skin and eyes. There are several forms of these hairs, used by different tarantulas against particular predators or parasites. Some are barbed, and dig deeper into the skin as they are scratched. A few species incorporate their hairs into a nest- and egg sac-defense strategy, weaving the hairs into silk to keep parasitic flies and wasps at bay.

Old World species do not have these specialized hairs, but often use aggressive threat displays to chase away enemies. These spiders often back up bright warning colors, threatening postures, and rasping hair rattling with an eager willingness to bite.

Fried Tarantulas, Cambodia
CC: Spotter_NL (Flickr)

Cobalt Blue Tarantula (*Haplopelma lividum*)
© Rachel Barton

Tarantula, Malaysia
© Eric Herve

Beauty and Design in the World of Spiders

Funnel-Web Tarantulas
Dipluridae

Most diplurids are visually distinguishable from theraphosid mygalomorphs, as their spinnerets are readily apparent—often more than half as long as their abdomen. Overall size varies by species; some are very large spiders while others are diminutive. These spiders craft a silk retreat tube, sometimes taking over another animal's burrow, then spin a funnel-like web which emerges out into the surrounding vegetation or leaf litter.

One of the few diplurids in North America is the endangered spruce-fir moss spider, *Microhexura montivaga*. This spider survives in moist clumps of moss and other bryophytes within high-elevation spruce and fir forests. It can only be found in a few counties in Tennessee and North Carolina. These are tiny diplurids, with adults measuring less than a quarter inch in length. One of the primary threats to this species is habitat loss, as invasive pests destroy the spruce-fir forests. The loss of trees affects the moss, as temperature and humidity changes create microclimates harmful to the spider.

Jamaican *Ischnothele* diplurids are targeted by tiny kleptoparasitic spiders of the genus *Mysmenopsis*. Each species of *Ischnothele* has a different *Mysmenopsis* species living in its web. The kleptoparasites steal prey from the diplurids, as well as consuming any tiny captured insects the diplurids ignore. In order to keep from being chased away, these kleptoparasites must move cautiously on the web threads so as not to trigger a feeding attack. They only move faster when the host spider itself is moving or busy feeding. *Mysmenopsis* will approach while *Ischnothele* is ingesting a meal, trying to scavenge something for itself, but must remain wary of their host's aggression.

Venomous Funnel-Web Tarantulas
Hexathelidae

While diplurids are certainly capable of giving a nasty bite, hexathelid spiders are notorious for one particular species that can be deadly—the Sydney funnel-web spider, *Atrax robustus*. Bites from other species should certainly be taken seriously, but *A. robustus* is especially dangerous. There are even differences between the sexes—male Sydney funnel-webs are much more toxic to mammals. Fortunately, anti-venom is available within Australia for any naturally occurring envenomations.

Like diplurids, hexathalids create a silken retreat, but most, like *Hadronyche* species, don't create a long spun funnel. Instead, the funnel is short, and develops into irregularly connected strands of threads that act as trip-lines. The spider perches at the edge of its retreat with its legs touching these silk threads, waiting for an insect to brush one as it walks past. As soon as it detects the vibration, the spider charges to grab its prey.

Sydney Funnel-Web Spider
© James van den Broek

TRAPDOOR AND TURRET SPIDERS
CTENIZIDAE, ANTRODIAETIDAE, CYRTAUCHENIIDAE, MIGIDAE, BARYCHELIDAE, IDIOPIDAE

Trapdoor spiders are so-called because they close the entrance of their burrow with a hinged lid or other barrier, allowing them to keep it closed and hidden until prey wanders close enough to be snatched. Turret or collar-door spiders lack an obstruction on their burrow's opening. Rather, they build up the ground around the entrance into a small mound of soil and vegetation.

Trapdoor spiders may dig out single or branched tunnels, lining the sides with silk. Some species pack mud on the sides of the burrow, or use "saliva" to help plaster dirt to walls. They begin with a burrow wide enough for them to dig in, then create the trapdoor. Once they have the trapdoor in place, they can work securely to dig out the rest of their burrow. They use rake-like hairs on their chelicerae to loosen the soil. Side shafts may be created as escape routes. The bottom of a tunnel is used for feeding, molting, laying egg sacs, and rearing spiderlings.

Trapdoor Spiders, Australia
© Ulrike Bauer

While hunting, the trapdoor spider sits at the entrance of the burrow, just behind the hinged door. When it feels vibrations at the edge of the burrow, the spider pounces, usually while keeping its posterior within the burrow for quick retreat. In some species, vibration-detection is enhanced by trip-wire threads or blades of grass emanating from the burrow. The use of these sensors varies with substrate, as certain soil types are not conducive to carrying vibrations.

The trapdoor's hinge is placed on the most elevated side of the entrance, so that the door naturally falls shut when not propped open. Some species bevel the edges of the door, so that it lies flush against an outward flaring entrance, making it difficult for predators to pull up. When disturbed, the spider grasps the underside of the trapdoor with its claws or fangs and holds it shut.

If a predator manages to break through, the spider heads out any available side exits. If it is a single tube burrow, the spider retreats to the bottom of the burrow, which is usually wider than the rest of the tunnel. It then turns to face its enemy with legs outstretched and fangs ready to fight. The abdomen of one idiopid genus, *Galeosoma*, has a hardened posterior that it uses to plug itself safely away from predators. Facing away from the invader, it jams itself into a narrow portion of the burrow, showing only its tough rear shield. This gives the predator no way to attack the spider.

During dry seasons, burrows are sealed up from the inside. This prevents water loss from females and their spiderlings. Sealed burrows may also indicate that the spider is molting.

Not all trapdoor spiders live underground. Some are arboreal, living in trees or on the face of a cliff or boulder. A silk tube is spun into a depression in the bark or rock before being camouflaged with debris to blend with the background. These tubes usually have a trapdoor at either end, which provides both greater scope for prey capture and an additional direction for escape if disturbed. A few species, like the migid trapdoor spider *Poecilomigas basilleupi*, build single-trapdoor tubes descending into a tree cavity or depression in a stump.

Turret spiders, like one genus of cyrtaucheniid spiders in California, *Apomastus*, do not use a trapdoor. The coastal dune species, *Apomastus simus*, digs a burrow in the sand, lining it with a thick layer of silk. This silk tube extends from the ground, and is often built up with a layer of soil and vegetation. The burrow extension provides greater opportunity to detect vibrations of insects that wander past the entrance.

PURSE WEB SPIDERS
ATYPIDAE

The purse web spiders are small mygalomorphs often measuring less than an inch in length. Coloration in these spiders is sometimes quite striking. In North America, *Sphodros niger* has an attractive black sheen. *S. abboti* is also dark, but with a purplish abdomen. *S. bicolor* offsets its black body with bright red or yellow legs.

These spiders bury a long silk tube into the soil, often at the foot of a tree. Habitat taste differs by region, some being found in rich organic soil while others prefer sandy soil. The tube extends up a tree or under the grass or duff, with an open end that naturally flattens when not in use. The visible white silk may be camouflaged, intentionally or not, by a fine layer of dirt. This also helps strengthen the tube structurally.

The spider waits in ambush just under the ground within the tube. If an insect, millipede, or other small invertebrate encounters the tube, the spider jumps forward and thrusts its large fangs through the dense silk, holding its prey until the venom takes effect. When the prey stops struggling, the spider cuts a gash in the silk, pulls it inside, and carries it back underground for feeding. The silk tube will be repaired later, and the prey's remains will be tossed out the tube's entrance.

In the spring, males begin their wandering as they search for a mate. Males of certain species appear to be generic wasp or ant mimics, which may provide them with some protection from predators as they actively hunt for mates. When they locate a female's tube, they drum on the silk and wait for a welcoming response. The female lays her egg sac in the burrow, with spiderlings emerging in the summer. The young spiders often disperse by ballooning to new regions.

Atypus sp., Russia
© Alexander M. Omelko

Sphodros sp.
© Tom Coleman, University of Kentucky, ForestryImages.org

Jumping Spiders
Salticidae

Often colorful, occasionally cryptic, most jumping spiders are active foragers that chase, stalk, and ambush their prey. They display intelligence (problem solving and planning) and clearly exhibit marks of design in their physiology. There are over 4,000 described species of salticids. Their hallmark leap is a predatory pounce (or a bound from danger) of several inches. Some species leap almost fifty times their body length. Salticids exhibit every color in the rainbow, even iridescence, and many species have distinctive patterns crucial for intraspecific communication. They are found in numerous habitats, and are commonly met in houses and backyards.

Jumping spiders do not usually spin capture webs, but they do create nocturnal retreats. These can be found under rocks and deadwood, or attached to vegetation. Often these will be a silken tube or dome-like structure, and in many species the female's egg sac will be attached to it or woven within it. Depending on species and latitude, North American salticids mate from early spring to early summer. Females spin an egg sac and lay their eggs in the summer or fall. Most females lay several egg sacs over this period of time. Fertility and number of eggs per sac decrease with each successive batch. Within a week of the egg-laying, eggs begin to expand with the spiderlings' growth, breaking free of constricting silk around the egg mass. Details of the egg-bound spider's morphology can be noted within another week. After a few more days, the spider

Jumping Spider, *Phidippus* sp.
© Dawn M. Turner

uses its egg teeth, located on its chelicerae, to split the egg open and squeeze out of it. Over the next several days the spider develops its pigmentation. These spiderlings remain within the egg sac for over two weeks, going through at least one molt before scattering from the hatching-ground. Many mature spiders, particularly males, will not survive a winter, but most of their immature offspring overwinter successfully. Offspring go through multiple moltings during the warmer seasons before they reach maturity.

Vision in Jumping Spiders

Vision is the primary behavior-triggering stimulus for jumping spiders. While chemical, vibratory, and other environmental cues are important, optical cues are foundational for salticids' diverse and complex behaviors. Both predation and courtship rely on a jumping spider's ability to detect and correctly distinguish other creatures visually.

Jumping spiders have four pairs of eyes, with some variation among genera in arrangement on the spider's carapace. The largest pair is situated directly in front, facing forward. These are the principal eyes, also called the anterior medial eyes. The other three pairs are secondary eyes, and are situated in rows along the top of the carapace. The anterior lateral and posterior lateral are moderate-sized eyes found, respectively, before and behind the minute pair of posterior median eyes.

Jumping Spiders From Around the World
Top, down: © Jason Ng; © Andrew Williams; © Clarence Seet; © Hway Kiong Lim

Each of these eyes consists of the same basic components: cornea, fixed lens, retina, and vitreous body. The secondary eye capsules are cup-shaped depressions, while the primary eyes are held in elongated eye tubes. The differences in eye complexity allow the visual process to be divided between the primary and secondary eyes.

The secondary eyes serve as motion detectors. Each of the anterior lateral eyes has a forward-projecting visual field of about 55°, while the posterior lateral eyes perceive movement within a 130° field on both sides of the spider. These readily discern motion up to about ten inches away. Researchers noted that the salticid *Phidippus audax* is able to detect the outline of a patrolling dragonfly against a bright sky from about three meters away. Once movement is detected, the spider moves closer and orients itself so that its principle eyes can focus on the activity. The distance the spider is able to discern the identity of a moving object depends on the spider's size, but some adult salticids recognize other jumping spiders up to eight inches away.

The binocular-like principle eyes of a jumping spider are unique among invertebrates. Light passes through the image-magnifying corneal lens and travels down the eye tube to a narrowing concave pit that acts as a second lens, essentially creating a telephoto-lens optical system. The image is transmitted directly to the retina, which consists of four separate layers of light receptors. Because light passing through this telephoto-lens system is broken up into separate wavelengths, different colors come into focus on different retinal layers. This tiered retinal arrangement means that jumping spiders are able to distinguish green, blue, and ultraviolet colors with their principle eyes.

The last retinal layer's foveal region is designed with enough light receptors to focus with high acuity on the image, but not with so many that quantum interference would disrupt the image. Another remarkable feature is set up to allow objects at different distances to be seen in focus. Unlike human eyes, the jumping spider's lens cannot change shape to allow such focusing. Instead, these foveal receptors are positioned in a "staircase" fashion, so that they are fixed at different distances from the lens. The image passing through the telephoto-lens will always be in focus on some part of the foveal receptors. To ensure this, the eye tubes themselves move back and forth while the corneal lenses remain fixed on the object of interest, allowing the image to be swept across the staggered foveal receptors.

How well can jumping spiders see? The Old World salticid genus *Portia* has been studied extensively. Its visual acuity is ten times greater than that of a dragonfly's compound eye, and only five times less than a human eye. Being so small, a jumping spider's principle eye cannot quickly perceive wide expanses of detail in an object, but it appears able to process visual information in sections by moving its eye-tubes in methodical search patterns.

Smaller objects are discerned more quickly, with the jumping spider having to decide whether the object is potentially prey, a mate, a competitor, a predator, or of no real interest. While not every decision is based solely on visual data, certain recognizable characters immediately elicit stalking, aggression, or courtship. While not proven, it may be that these characters provoke a quick response because they match specific pattern configurations in the arrangement of foveal receptors.

Communication Displays in Jumping Spiders

The active communication displays for which jumping spiders are best known come in two categories: male-male agonistic displays and male-female courtship displays. Displays are primarily visual, requiring specific characteristics in size, shape, motion, and coloration. Recent studies with a tropical ornate salticid, *Cosmophasis umbratica*, suggest that even ultraviolet reflectance can be important in displays.

Aggressive or agonistic displays are territorial in nature, as two males approach each other, raising their bodies, opening their chelicerae, lifting their forelegs, and extending their palps. This allows each male to test the size of the other, giving a smaller male opportunity to retreat without injury. Similar-sized males advance on each other, moving their appendages and abdomen in species-specific

Maratus sp., Australia
© Farhan Bokhari

Chinoscopus sp., Panama/Costa Rica
© Dr. Arthur Anker, Smithsonian Tropical Research Institute,
Panama City, Panama

patterns, holding their bodies in particular postures, and eventually moving close enough to touch each other. This leads to embracing, pushing, and grappling. The physical fighting continues until one spider retreats, or is injured or killed.

Courtship displays are triggered in male jumping spiders when they gain visual contact with a female. The female remains still as it scans the male, assessing whether it is prey. The male begins species-specific courtship movements, following a zig-zag path towards the female. Any other path could trigger a spontaneous predatory attack by the female. If she is receptive, the male waves, thumps, and skitters his way to her until he is close enough to mount her.

Displays sometimes include a vibratory component, though one that is not always perceptible to human ears. Some male jumping spiders scrape a file on their carapace or palp, some vibrate their abdomens to produce a sonic buzz, and others drum parts of their body on the ground. Both ground-borne and air-borne signals are produced.

The male of one species of jumping spider, *Maevia inclemens*, has two different physical forms—"tufted" and "gray." The first has a solid black body, white legs, and tufts of setae on the front of its cephalothorax. The second morph has a black-and-white striped body, orange pedipalps, and no tufts. Not only do these two morphs look different, they each have a distinctive courtship behavior. The tufted morph "stilts" about 9 cm from the female, raising its body high off the ground while waving its front pair of legs and rocking its body back and forth. The gray morph approaches the female in a crouch, coming up to about 3 cm away, and edges side to side with its first two pairs of legs pointed forward like a triangle. Because these different displays occur at different distances, each display takes up about the same space in the female's visual area. This allows two distinctly different male forms to trigger courtship receptivity in the same female.

Prey for Jumping Spiders

Flies are common prey for jumping spiders, but larger salticids overpower insects as large as a dragonfly. Grasshoppers, lepidopterans, and other spiders also join the lunch menu. Beetles are less common, as wing covers and dorsal shields can be difficult to penetrate, and some beetles use noxious secretions against predators. Prey are typically stalked or ambushed while the jumping spider wanders its territory. Males are usually most active, searching for both a mate and a meal.

Jumping spiders react with genetically programmed prey-specific instincts. Flies are stalked from behind to just within leaping distance, while caterpillars are approached much closer before attacking the head. This versatility in predatory behaviors relies on visual input and recognition. *Phidippus audax* readily differentiates between two soft-bodied yellow beetles by the number and pattern of dark spots on their dorsum. The spotted cucumber beetle is a favored meal, but the soldier beetle is quickly avoided.

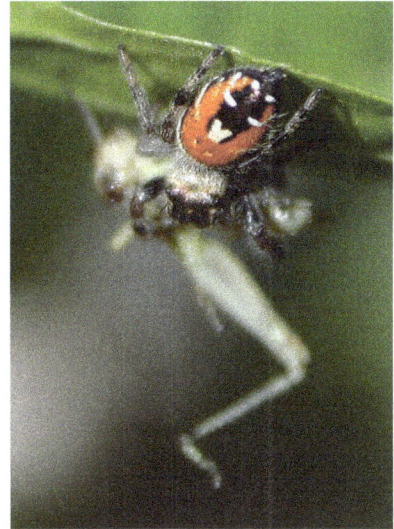

Jumping Spider with Prey
© Dawn M. Turner

Salticids that ambush dangerous or larger prey use strategic tactics to keep the upper hand. With struggling prey, they drop on their dragline into mid-air, holding their prey until it stops moving. This keeps the prey from leveraging against the spider as it tries to escape. The same tactic also hinders social ant prey from gathering reinforcements through their alarm pheromones. The chemical draws fellow colony members, but they are unable to descend the spider's dragline.

Several salticids are specialist ant predators. The vast number of worker ants in a multitude of habitats provides a constant available food source, but ants are not passive creatures. Stinging, biting ants require strategies for overpowering and pinning, or holding dangerous elements out of reach. Ant-eating salticids vary their predatory behavior depending on size, and learn to avoid ant species with particularly repugnant secretions. Ants produce chemical trails for their own benefit, and some salticids detect these chemical signatures, provoking a predatory readiness in the jumping spider that increases its desire to hunt prey.

Bold Jumping Spider (*Phidippus audax*)
© Cathy Keifer

Jumping Spider, Brazil
CC: Anderson Mancini

Beauty and Design in the World of Spiders

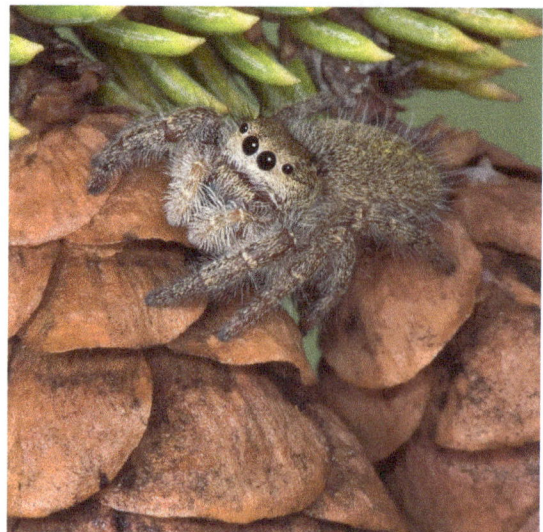

Top Left, Clockwise: *Viciria* sp., Singapore (© Dr. Arthur Anker, Smithsonian Tropical Research Institute, Panama City, Panama); *Corythalia* sp., Costa Rica/Panama (© Dr. Arthur Anker, STRI); *Phidippus audax* (© Howard Cheek); *Phidippus princeps* (© Cathy Keifer); *Maevia inclemens* (© Howard Cheek).

Portia: The Ultimate Mimic

One of the strangest, and most-studied, salticids is the Old World genus *Portia*. This genus of spider-specialists has developed a wide range of methods to lure other spiders to their death. Some spin a web, allow insects to get tangled in them, and then wait until klepto-parasitic spiders come calling. *Portia* mimics pieces of debris, pulling in its legs tight to keep its outline from scaring away prey. As the kleptoparasitic spiders try to steal the insects from the web, they end up a meal themselves. Other *Portia* invade orb-webs, luring the resident spider into reach by plucking and thumping on the web strands in imitation of a captured insect. Such behavior is called aggressive mimicry. Most orb-weavers have poor eyesight, so they must rely on web vibrations to determine when prey has been captured by the sticky threads.

Portia, from Singapore
© Melvyn Yeo

Several tactics point to a level of intelligence in these jumping spiders. While some behavior is genetically programmed, *Portia* is clearly capable of trial-and-error learning, trying different tactics until it gets the correct response. At times it lures the spider in slowly; other times it plucks monotonously to lull the spider while it creeps up on it. Some web-spinners are large predators, so *Portia* has to make certain their dangerous parts are out of the way. As it creates these vibrations, *Portia* manipulates the other spider into moving into just the right position for an attack.

Portia also takes advantage of wind blowing on the spider web, which creates a vibrational smokescreen for the jumping spider to quickly move closer to its victim. *Portia* discerns correct paths to its prey, even if the path requires a detour away from it. These detours may take it out of sight of the victim for long periods of time, but *Portia* is able to focus on its task. Even when hunting on the outskirts of social spider webs, *Portia* focuses on a particular target rather than just making generic signals in hopes that one spider out of many will fall for it.

Not all of *Portia*'s victims are orb-weavers. Other jumping spiders are also hunted. When *Portia* encounters the silken retreat of another salticid species, it strums the silk carefully, trying to lure out any resident spider. The Queensland, Australia, population of *Portia fimbriata* has developed a novel technique, cryptic stalking. When it encounters another jumping spider out in the open, this species freezes and pulls in its palps and legs, making it look like just a bit of detritus. It then moves very slowly, with an irregular jerky gait, each leg moving independently, until it is close enough to pounce. If its victim detects movement and orients toward *Portia*, the predator stops again and waits until the other spider turns away. *Portia* only uses this tactic against other jumping spiders; if it stalks other prey, it does so without the extra care not to be recognized as a salticid.

Portia itself sometimes needs to take care. The female of a Kenyan species, *Portia schultzi*, initiates a courtship drumming just to lure in males to eat, though she might let him mate first. Immature females which can't yet breed will also lure in males, and even use chemical signals that mimic breeding pheromones to do so.

Mimicry in Other Jumping Spiders

Among the salticids are several species which exhibit mimicry. In some species, the mimicry is strictly morphological, while in others it includes chemical and behavioral disguises. Pseudoscorpions, ants, and beetles are common models for salticid mimicry. One Indian species, *Rhene danielli*, shares a remarkable resemblance to a paper wasp, while a Borneo salticid, *Orsima formica*, mimics a generalized insect in reverse. Its abdomen is held high, with long spinnerets appearing like antennae and mandibles. Mimicry among salticids provides protection against spider predators, allows the spider to closely approach potential prey, or, in the case of some ant mimics, allows the spider to live safely within an ant colony.

Ant-Mimicking Jumping Spider with Ants
© Bill Beatty

Myrmarachne maxillosa, Singapore
© Raynard Heah

Because ants are formidable prey for invertebrate predators, using an arsenal of stings, poison glands, and crushing mandibles, they make excellent models for mimicry. Some spiders, like *Synemosyna americana*, developed elongated abdomens and patterns which suggest additional body segments. Often, the front legs are held off the ground to mimic antennae, and the spider moves in an erratic ant-like manner. Ant-mimicry, or myrmecomorphy, can become quite complex, as some spiders exhibit transformational mimicry (their ant species model changes throughout their life cycle, as the spider grows from juvenile to adult) or polymorphic mimicry (different sexes or different color morphs of the adult jumping spiders mimic different ant species). Some ant-mimics live in close association with an ant colony, enhancing protection from predators and parasites. This can be challenging for the spider, which must keep from being seen as a threat by the colony. One Australian salticid, *Cosmophasis bitaeniata*, lives in shrubs along with the aggressive and territorial *Oecophylla* green tree ants. It steals and eats ant larvae, so it gains colony-specific chemical extracts that are synthesized into the spider's body. Thischemical mask fools the worker ants.

Strangely enough, while these jumping spiders have transformed morphologically to look like something other than a spider, some tephritid flies mimic jumping spiders to their own advantage. They use leg-like striations on their wings to confound predatory salticids. When approached by a jumping spider, these flies wave and twitch their banded wings to the side of their bodies, imitating the aggressive leg-waving behavior of salticids. This causes the spider to stop and answer with its own display, giving the fly time to escape.

CRAB SPIDERS
THOMISIDAE

Some spiders are well-camouflaged ambush predators, often with anatomical modifications for this particular lifestyle. Crab and lynx spiders, for example, have raptorial front legs that are enlarged for striking and grasping prey.

Thomisids chiefly surprise prey from flowers, foliage, or blades of grass, though some species stalk through leaf litter. Specific plants, such as goldenrod, clover, pasture rose, sunflowers, or milkweed, may be preferred at different times as ambush points. The spiders move to new locations when pollination cycles change—pollen runs low or seasonal flowers bloom. Common prey includes butterflies, bees, wasps, beetles, and flies. While primarily day hunters, at least one species, *Misumena vatia*, will wait in ambush on milkweed at night in order to capture moths.

Crab spiders are able to distinguish between flowers that are producing pollen and those that are not. They choose their ambush position to correspond to those flowers that are more likely to attract pollinating insects.

Crab Spider Positioned for Ambush, Spain
© I. M. Verdu

This ability to determine profitable ambush sites may be through chemical or visual cues. If a chosen ambush site is not productive within a few hours, the spider moves on to another flower.

Just as a successful hunt by lions will draw scavengers like hyenas and jackals, so will a successfully hunting spider draw other species wanting to share a meal. Kleptoparasitic spiders live on the edges of orb-webs waiting to pluck unguarded or unwanted prey. Crab spiders don't spin a web, but their predation draws the attention of wandering harvestmen. Harvestmen are spider-like, but can be distinguished by the lack of a pedicel between their cephalothorax and abdomen. Harvestmen will feed on a crab spider's discarded insect remains, but sometimes will try to steal prey directly from the spider's grasp.

One Asian crab spider, *Misumenops nepenthicola*, lives along the inside rim of insectivorous pitcher plants, *Nepenthes* spp. It feeds on insects that are attracted to the plant's glandular secretions. It also feeds on mosquito larvae that inhabit the pitcher plant's enzymatic liquid. In order to do this without becoming a meal, the spider affixes a safety thread at the top of the pitcher before it dives in. Without this thread, the spider will not be able to climb up the slick walls and out of the digestive solution. This spider also uses the pitcher plant's liquid to escape potential predators, being able to survive for long periods under the fluid's surface.

Crab Spider Capturing Beetle, Ecuador
© Dr. Arthur Anker, Smithsonian Tropical Research Institute, Panama City, Panama

Misumenops sp., Slovenia
© Ziga Camernik

Thomisus onustus
© David Acosta Allely

Crab Spider, India
CC: Subramanian Kabilan

Most crab spiders are cryptic, though the camouflage may be colorful itself. In several species (e.g., *Misumena vatia*, *Misumenoides formosipes*, and *Misumenops asperatus*), the spider's skin is transparent, so they appear white in normal light. When they sit on a yellow flower, however, the light that reflects from the flower's petals stimulates the production of a yellow pigment in the hypoderm layer of the spider's skin. The longer the spider stays on the flower, the more intense the change in its yellow hue.

Coloration that appears cryptic to human eyes may not deceive the eyes of other animals, however. Predatory birds and potential insect prey have vision systems that include both chromatic (with ultraviolet ranges) and achromatic vision. Crab spiders tested with such systems do compensate for differing visual ranges, blending in with at least portions of the chosen flower. One Australian crab spider, however, takes advantage of bee's visual preferences for eye-catching floral patterns. *Thomisus spectabilis* appears to blend in with a white daisy, at least to human eyes. The white spider reflects ultraviolet light in high contrast with the flower itself, though, a configuration that attracts bees and other pollinating insects.

Small juvenile spiders may have abdomens that change color depending on what they have recently eaten. As they grow, their integument blocks dietary pigments from affecting the spider's external coloration.

Thomisus onustus
© Smag / Fotolia

Misumena vatia
© Howard Cheek

Synaema globosum, Spain
© Ismael Montero Verdu

Besides color camouflage, some crab spiders practice mimicry. *Misumenoides formosipes* nestles among a flower's petals, sticking its front pairs of legs in the air. The legs resemble the flower's thin sepals. Other crab spiders, like *Phrynarachne* spp., mimic bird droppings in shape and color.

When they are ready to lay their eggs, female crab spiders choose a leaf, and bend the tip of it under, fastening it in place with silk. The spider hangs upside-down and lays the eggs. Often, the spider will spin threads to a second leaf underneath the eggs, pulling the two leaves together, providing stability and a vegetative hideaway for the female. This allows her to guard the eggs while keeping out of sight of enemies. Eggs are guarded from predators like ichneumonid wasps. The wasps attack by laying a larva on the eggs, which feeds on the eggs as it grows, often consuming the entire clutch before pupating.

Spiderlings grow throughout the summer, molting several times. These overwinter as immature spiders, and do not molt to their adult stage until the following spring. Once mature, females are large enough to tackle the biggest prey (like bumblebees), which spurs a faster increase in body size. Larger females produce more eggs. Males do not get nearly as big, and don't feed as often. Instead, they spend much of their time searching for mates.

Males in some species, like *Misumenoides formosipes*, guard female

Juvenile Crab Spider with Abdominal
Pigmentation
© MDB / Fotolia

Ant-Mimicking Crab Spider (*Amyciaea* sp.),
Hyderabad, India
© Maneesh Kaul

Phrynarachne sp., Madagascar
© John Roff, Hilton College, South Africa

spiders that are about to molt into reproductive maturity. They chase away other males, often engaging in fierce combat. In *Misumena vatia*, there are usually more females than males born in a clutch of eggs, so male rivalry is not necessary.

One Australian thomisid, *Diaea socialis*, is semi-social in nature. After a female lays her eggs, the spiderlings continue to live with her as they mature. They form a colonial nest by attaching eucalyptus leaves together, share prey with each other, and mate with their siblings. After mating, females will go off to form their own families.

Lynx Spiders
Oxyopidae

Lynx spiders stalk among the foliage, with coloration matching the vegetation. When close, they attack and subdue prey with their venom. Oxyopids are generalists, and significant predators of pest insects. A few invade orb-weaver webs and cobwebs to attack the resident spider or steal entangled prey. Only in one genus, *Tapinillus*, do lynx spiders build webs for capturing prey.

Male green lynx spiders, *Peucetia viridans*, recognize mates visually. Once a male meets a female, he vibrates his abdomen, waves his forelegs, and drums his palps. He continues this as he slowly approaches. When he is within reach, he touches her forelegs to determine her readiness. If she doesn't attack, and is willing to be courted, she raises her legs for mutual stroking of each other's legs and patella. After a period of tactile courtship, the female jumps from the vegetation and hangs in mid-air from her dragline. The male watches from the edge of the foliage, then twirls the thread until she faces the right direction. He drops on his own dragline and mates with her. When finished, they crawl back up, but may repeat the process several times, jumping off the leaves time and again.

Lynx spiders lay eggs for only one season, though they may produce more than one egg sac. Females vary in size, but all produce eggs optimized for survival. While smaller females have fewer eggs, these eggs are the same quality as those produced by larger females. This ensures that each egg's energy content allows it to develop and hatch successfully.

Female lynx spiders guard eggs aggressively. The green lynx spider sprays a fine stream of venom at some attackers. While more noxious than dangerous, care should be taken not to get this into an eye or on a mucous membrane.

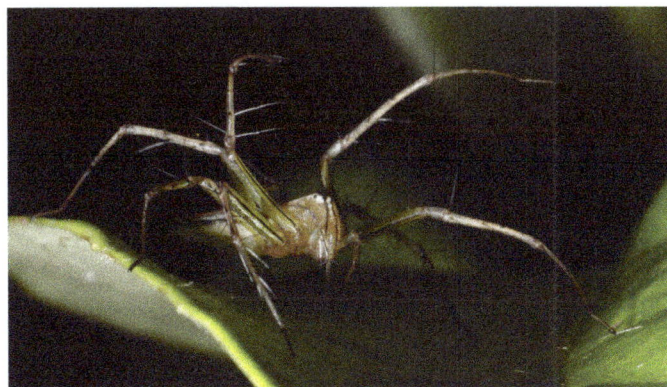

Lynx Spider, South Carolina
© Steven Faucette

Lynx Spider, Singapore
© Sim Kay Seng

Lynx Spider, Australia
© Douglas Stetner

Lynx Spider, Sri Lanka
© Keith Blom

Lynx Spider, India
© Geo Paul

Monkey-Faced Spider,
Tapponia sp.
© Yixiong Cai

Lynx Spider, Madagascar
© John Roff, Hilton College,
South Africa

Velvet Spiders and Kin
Eresidae

Eresids are variable in coloration and habit. Adults are covered in velvet-like hair. Males of several Eurasian ladybird spiders (*Eresus cinnaberinus*, *E. sandaliatus*, etc.) are among the most beautiful arachnids. While the females are velvet black, the males have a black cephalothorax, striking white banding on their legs, and a bright red or orange abdomen with four distinct black spots.

Ladybird spiders construct a silk-lined burrow among the vegetation, installing radiating trip-lines to warn of passing prey. Similar ambush methods are used by other eresids. The Namib desert ant-eating spider, *Seothyra henscheli*, must contend with surface sand temperatures that could easily kill it. It spins a small prey-capture web, but spends its time waiting underground in its burrow. If an ant entangles itself in the web during the day, the spider takes several trips back and forth as it envenomates its prey, untangles the ant from the web, and drags it into the burrow. The surface is too hot for one trip, and the spider must cool off several times before the job is finished.

Several species of the eresid genus *Stegodyphus* are social spiders. Most of these spin silk passage-riddled nests in scrub brush that allow them to wait in ambush near outside capture webs. These "family spiders" cooperate in catching prey, fighting off enemies, maintaining the nest, and caring for spiderlings.

Maternal care in solitary eresids often becomes extreme. The mother captures prey and returns to regurgitate nutrients for her offspring. When the spiderlings have begun to grow, they turn on their mother and eat her. In the solitary *Stegodyphus lineatus*, male spiders practice infanticide, destroying any egg sacs they find with a potential mate. To counter this behavior, early-mating females put off laying their eggs until later in the season, reducing the chance of wandering males coming across their egg sac.

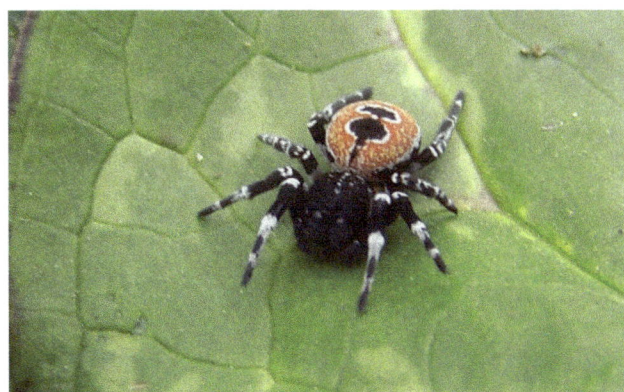

Eresus cinnaberinus, Male, Russia
© Sergey Pristyazhnyuk

Eresus sp., Female, Portugal
© Sérgio Silva Henriques

Eresus sp., Male, Portugal
© Sérgio Silva Henriques

WATER SPIDERS
CYBAEIDAE

The European water spider, or diving bell spider, *Argyroneta aquatica*, is almost fully aquatic. It spins a bell-shaped silk nest among the aquatic vegetation in ponds and other slow moving freshwater. The spider fills the nest with air bubbles carried from the surface. The spider's legs and abdomen are covered with hairs which trap air bubbles when it submerges. As required, the spider replenishes the air supply with more bubbles.

This silk nest serves as a refuge for the spider to rest, feed, molt, court, mate, and lay eggs. From its underwater retreat, females watch for tadpoles, small fish, and invertebrates to come close enough for ambush. Males stalk along the bottom of the pond at night. Unlike most spiders, the male water spider is larger than the female. When mature, the male water spider spins a nest, attaching it to a female's silk dome. He then chews through the wall to gain access to the female. *Woodcut from Furneaux (1911).*

Argyroneta aquatica
© Josef Hlasek

WOLF SPIDERS
LYCOSIDAE

Wolf spiders are generalist predators, pursuing just about any small species they run across, whether insect, spider, or small vertebrate. The largest wolf spiders are equipped to tackle mice and lizards. Wolf spiders will sometimes scavenge on dead prey, as well.

A number of lycosids live for several years. In some species, like the large North American *Hogna carolinensis*, spiders may not even mature until their third year. Females sometimes live another year after producing young, while males die soon after mating.

Courtship in wolf spiders is multimodal, comprised of visual and vibratory elements. Males jerk, tap, and wave their legs. Their appendages may have dark patterns and hair tufts to help attract the female's interest. Some species also pound the ground with their abdomen, drum their palps, or stridulate by rasping together two specialized components of their body, a "file" and a "scraper." This stridulation creates a buzzing sound. By using both visual effects and vibrations, males are able to court effectively during the day and at night. Some of these sounds are audible to human ears from several meters away. *Lycosa gulosa* has been called the "purring spider," because of its leaf-litter drumming.

Some lycosids have more than one clutch of eggs in a single season. While a few wolf spiders (*Arctosa* species, for example) protect their egg sacs in a silk burrow, many females carry their egg sac underneath their abdomen. This active care allows the spiders to keep the eggs safe from mold and parasites. It also gives the spider a way to regulate humidity and temperature. Some females will "sun" their eggs at the surface of their burrow, promoting embryological development.

This "free ride" inside the egg sac doesn't end when the spiderlings hatch. After emerging from the cocoon, the spiderlings flock to the top of the mother's abdomen and grasp specially modified hairs. These knob-tipped hairs have small curved spines running along their shaft, so the spiderlings can hang on with their claws as the mother travels. After the spiderlings learn to use their drag lines and fend for themselves, they leave their mother and head off on their own.

Lycosids do not spin a silken retreat as many spiders do. Instead, most just dig a burrow in the ground. They start by spinning a small round silk

Carolina Wolf Spider, Feeding on a Stolen Eggsac
© David Bronson Glover

Wolf Spider, Russia
© Kristina Rogova

Wolf Spider on the Beach
© Steve Shoup

Beauty and Design in the World of Spiders

48

mat on suitable soil, then push it aside along with any sand or dirt that sticks to it. They do this over and over until they have a suitably deep burrow. Larger wolf spiders may not bother to expend the energy for this. Instead, they'll take over burrows dug by other animals.

Wolf spiders wander far afield from their burrow in search of prey, but are able to accurately return without difficulty. In several lycosids, this homing trait stems from their anterior median eyes' ability to detect polarized light. By analyzing patterns of polarized light in the sky, the wolf spiders orient themselves in the direction of their underground retreat.

There are a few lycosid groups that don't follow the established patterns. Mid-summer, *Gladicosa pulchra* shifts habitats, moving from the ground to an arboreal lifestyle, possibly to reduce confrontation with other terrestrial lycosids. In the funnel-web wolf spiders, *Sosippus* spp., the females spin a funnel-like web extending from their burrow. Spiderlings remain in their mother's retreat for up to five months, feeding on prey the female spider drags in from its web.

Another unusual group of lycosids are the burrowing wolf spiders (*Geolycosa* spp.). Many of these spiders dig a vertical burrow in which they remain for most of their lives. They don't actively search for prey, but rather ambush insects that wander past the mouth of the spider's burrow. Courtship has not been extensively observed in this group, but it appears that adult males detect when a female is about to molt into maturity through chemicals in her burrow silk, and camp out at her burrow's entrance until she is ready to mate.

Wolf Spider Warming Eggsac at Burrow's Entrance
© Steffen Foerster Photography

Wolf Spider and Spiderlings, Texas
CC: Clinton and Charles Robertson

Geolycosa burrow
© Whitney Cranshaw, Colorado State Univ., ForestryImages.org

Arment | Eight-Legged Marvels

Kaua'i Cave Wolf Spider, *Adelocosa anops*
© Gordon Smith (both)

This Hawaiian wolf spider, found only in certain Kaua'i lava tubes, is an endangered species. It is blind, having lost its eyes. This cave species also has a lower rate of reproduction than found in typical wolf spiders, and produces 30 or fewer large offspring in a brood.

Eyelessness in cave species is not uncommon, resulting from genetic mutations interfering with eye development. Natural selection favors the loss, which protects the animals from infection and disease through inadvertent injury. The phenomenon shows how philosophy affects the interpretation of facts. One person points to eyelessness as evolution in action, where genetic mutation and natural selection create a species that is best fit for a specialized environment. Another person notes that loss of genetic information, even advantageous loss, is the wrong mechanism to promote macroevolution. After all, the same facts fit the concept of design, or speciation from discontinuous groupings, not just common descent.

FISHING AND NURSERY-WEB SPIDERS
PISAURIDAE

Most pisaurids are semi-aquatic, spending much of their time hunting on aquatic vegetation at the edges of streams, creeks, and ponds, or on the water itself. They catch and eat a variety of small aquatic invertebrates, while larger species capture small fish, tadpoles, and frogs. Fishing spiders successfully catch prey more than four times heavier than themselves. Their coloration blends in with the surrounding vegetation, and if disturbed by predators, fishing spiders are able to submerge and hide underwater for long periods of time.

Fishing spiders are also called nursery web spiders because they spin a silk tent in which they place their late-stage egg sac. This "nursery" holds the emerging spiderlings until their second molt. Juveniles often take some time to mature. In northern populations, *Dolomedes triton* may go through two or more winters before reaching adulthood.

Courtship in one species of *Dolomedes* begins as the male waves its legs and thrashes the water in response to encountering sex pheromones. As the male approaches the female, he taps his legs as the female waves hers. This courtship ritual can go on for half an hour before they finally mate. With *Pisaura mirabilis*, the male goes a step further. He catches a small insect, and co-coons it with silk. He presents it to the female, which she grabs and starts to eat, giving the male opportunity to mate. The gift acts as a "sensory trap," in that the more it looks like an egg sac (the whiter and rounder it is), the more likely the female is to accept it. Females of this species typically carry their egg sacs in their chelicerae, providing a maternal instinct for the male to exploit. So long as the female has not yet produced an egg sac, her impulse immediately turns to feeding on the gift.

Fishing spiders can walk, sail, and gallop on water. Their bodies are functionally hydrophobic, causing water to bead when it comes into contact with the spider's skin. This not only allows them to remain on top of the water without breaking surface tension, but even when they dive, they stay dry while submerged. When the spider remains motionless, this hydrophobicity creates dimples in the water's surface under each

Dolomedes sp.
© Michael Pettigrew

Dolomedes minor, New Zealand
CC: Bryce McQuillan

Pisaura mirabilis and Prey
© Jorge Almeida

leg so that there is very little contact between foot and water. To move forward, the spiders must push pairs of their legs backwards against these dimples, effectively "rowing" across the water even without direct friction of their legs on the water's surface. The speed of this rowing is size-dependent. Smaller spiders row faster.

Smaller fishing spiders can also take advantage of wind, by elevating themselves on their legs, high off the surface of the water. Their hydrophobic legs provide very little friction on the water's surface, so they sail along, albeit with little control over direction. They can stop sailing, however, by changing body posture.

Fishing spiders detect prey through vibrations. They can accurately determine the center of a ring of concentric prey-generated waves from over seven inches away. Often, they sit on floating vegetation or at the water's edge, with pairs of their legs sitting on the water so that they detect the motion of aquatic prey. Rather than rowing, which cannot be done quickly, fishing spiders "gallop" towards their prey. They leap onto the water, rapidly retracting their legs after pushing into the surface, allowing the spider to reach its prey with the least possible contact with the water's surface. This "frictionless" locomotion features one further trait—brakes. Without a way to stop themselves, the spiders would shoot past their intended prey. So, as spiders begin their gallop, they attach a dragline to something solid, and hold the thread in one of their claws. When they reach the prey, the thread is grasped tight, stopping the spider just as it grabs its meal.

Dolomedes sp., Lurking at Water's Edge
© John Gayler

Dolomedes sp., Carrying Eggsac
© Oliver Arnold

Dolomedes triton, Six-Spotted Fishing Spider
© Lew Scharpf

Beauty and Design in the World of Spiders

A Spider's Design?

An Ingenious Spider. (1902)

Editor, *Popular Science News*:

Some years ago while crossing a pasture field in western New York, my attention was arrested by what seemed to be a large, variegated flower of about three inches in diameter in a tuft of grass. The slanting rays of an afternoon sun, on a bright August day, caused it to sparkle with a variety of tints, making it an attractive flower to the eye.

On close inspection I was astonished to find it an artificial flower, and the artificer and inventor a spider of the funnel web weavers, the Agalenidae family.

This spider had woven a regular funnel web three inches in diameter, and four inches deep, with an aperture at the bottom of about five-eighths of an inch in diameter. The petals of the flower were made of the wings of insects she had entrapped in her web and eaten; and undoubtedly the idea of converting the funnel into a flower was an afterthought, as all the parts of the flower showed plainly that they had been added to the original structure. There were in the outer circle of artificial petals seven wings of *Piera rapis* (cabbage fly). The second circle of petals was in continuation of the first, and made out of the bright, and redder wings of other insects. There was a third circle of petals of still smaller wings, but including a greater variety of colors than either of the other two circles of petals. The second circle of wings used for petals included eleven; and the third circle contained seventeen wings. The arrangement of every wing was such as to secure the best exposure of its colors, and that made a proper blending of colors to stimulate a flower almost as perfectly as human skill could do it.

A careful and thorough study of the structure fully convinced me that it was in every one of its many details a work of intelligent design. There was not discoverable a single feature of accident about it. The wings used for petals were arranged according to size first, and next their colors. They were securely bound in their positions by lines of web that passed around them and were made fast to the original web.

The spider, the female, remained at the bottom of the web. She could not be lured to the top of the web except by dropping some insect upon it, when she would come up and securely fasten it by passing lines of web around the insect and fastening them to the structure of the funnel below the petals of the flower. I dropped a number of insects into her web, and each time she repeated the operation of securing them; but did not eat any of them while I watched her. After she made one of the insects fast she would immediately retire to the broom of the web.

Directly under the opening of the web, at the bottom, there was a large quantity of the remains of insects upon which she had fed. Among the lot I discovered many times more wings of butterflies and other insects than she had used in making her artificial flower. She had evidently selected the ones she used with care in respect to their size and colors, for these features were prominent in the structure.

After a long and minute study of that artificial flower web, I left it with the conviction that it was an ingeniously designed means of the more readily luring the prey into the web; and much reflection upon it since, and even the failure to have ever found another such instance, have not shaken my conviction of it being a work of design.

(Rev.) A. N. Somers.

FUNNEL AND GRASS WEAVERS
AGELENIDAE

Often discovered running along the floor in houses or the rafters of barns, agelenids are most commonly associated with silken open-tube retreats expanding into small sheet webs laying on the grass or among ground-covering foliage. Some species prefer well-forested areas with damp organic soil, where they spin their tubes under or within rotting wood or at the base of tree trunks.

Agelenids ambush prey that encounter their funnel webs, but also hunt in the grass or leaf litter. Most funnel weavers are small, preying on crickets, roaches, springtails, and other spiders. There are larger species, like the introduced giant house spider, *Tegenaria duellica*, which, though harmless to humans, tackle larger invertebrates.

Most agelenids are solitary spiders, but the West African *Agelenopsis consociata* is social. Communities sometimes consist of well over 1,000 spiders. Cooperative living reduces individual energy requirements for a challenging tropical environment. In larger colonies, networks of silken sheets and funnel webs bridge multiple nests. The spiders cooperate in capturing large prey, as well as in defending nests from predators. Certain lepidopteran larvae, beetles, and ants wander these nests, allowed to eat insect remains or other refuse, keeping the nests clean.

In North America, there has been great interest in the expanding range of the European hobo spider, *Tegenaria agrestis*. First reported in Washington state in the 1930s, hobo spiders have moved into other parts of the Pacific Northwest. This spider's bite has been reported to destroy tissue. Prior to the expansion of the hobo spider's range in the Pacific Northwest, such bites were blamed on brown recluses, despite little, if any, evidence that populations of brown recluse exist in that region.

Tegenaria sp.
© Alistair Scott

Funnel Weaver's Webbing
© Lew Scharpf

Agelenopsis sp., Spinnerets Prominent
© Photography by Marcy Kellar

The most unusual aspect of hobo spider bites in North America is that the spider is medically insignificant throughout its native range in Europe. Toxicological testing shows that the North American population has no significant difference in venom composition from a population in the United Kingdom. It seems probable that ulcerating wounds blamed on alleged hobo spiders are often based on other medical conditions, just as brown recluses are often mistakenly singled out. Very often, the suspected spider is not even seen, or is only seen in the house rather than actually inflicting the bite. A wide range of hairy wandering spiders are misidentified as hobo spiders, and even within the agelenids, identification is tricky. Microscopic examination is required to conclusively distinguish hobo spiders from the inoffensive giant house spider, *Tegenaria duellica*.

Agelenid Waiting in Ambush, Germany
© Wolfgang Staib

Funnel Weaver at Home
© Michael Pettigrew

Agelenid Carrying Eggsac
© Beth van Trees

Sac Spiders
Anyphaenidae, Clubionidae, Miturgidae

In the field, sac spiders can be found under loose bark, in foliage, or under leaf litter. A few myrmecomorph species will be found with ant colonies. Sac spiders spin a silk tube or sac for their daytime retreat. Females eventually turn these sacs into a nursery, in which juveniles create their own moulting chambers. The microenvironment within the silk tube provides the right temperature and humidity to prevent dessication and provide safe overwintering. Both juveniles and adults overwinter, and maturation rates vary even within the same species.

At night, sac spiders search for food. They rely on vibrations to lead them to prey, although some have been recorded scavenging on dead insects. Sac spiders track down hidden larvae by their vibrations, moving slowly along the surface of a plant while the larvae feed underneath. The South American myrmecomorph *Corinna vertebrata* mimics worker ants in both form and gait, waiting until a solitary ant gets close enough before ambushing it for a meal.

More than one egg sac may be laid in a season. Female sac spiders often remain with their eggs in the silken nest. One reason they guard their eggs is that neighboring female sac spiders feed on them. The cannibalistic behavior found in one species of *Clubiona* was limited to females that had their own egg sacs. Males and non-maternal females never tried to feed on their neighbor's eggsacs.

Cheiracanthium species are agriculturally important, being active predators of invertebrate pests. They are commonly found in houses, and are a leading culprit in accidental spider bites. Because their venom is cytotoxic, sometimes causing minor necrosis, their bites are often misdiagnosed as brown recluse bites. These spiders are not life-threatening, however, and only bite if they are trapped against skin or deliberately agitated.

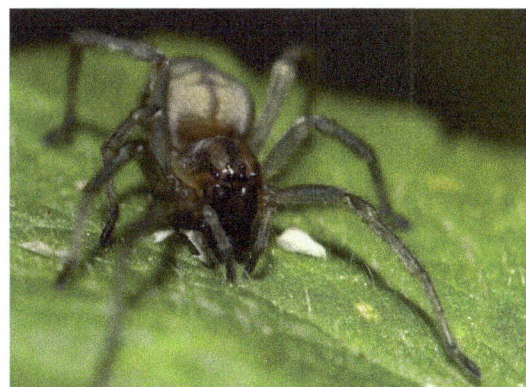

Ant-Mimicking Clubionid Spider
© Dr. Arthur Anker, Smithsonian Tropical Research Institute, Panama City, Panama

Yellow Sac Spider, *Cheiracanthium* sp.
© Cristian Draghici

Cheiracanthium sp., Kazakhstan
© Viktor Glupov

HUNTSMAN SPIDERS
SPARASSIDAE

Typical huntsman spiders resemble flattened wolf spiders that quickly move crab-like in any direction. They are well camouflaged for their habitat preference, whether dry ground, tree bark, cave walls, or vegetation. While these spiders are mostly tropical, a few species of *Olios* inhabit the American southwest. The banana spider, *Heteropoda venatoria*, has been introduced into several states by hitching a ride on imported produce. Exotic introductions are not uncommon in invertebrates, and one other large cursorial hunting spider, the Old World species *Zoropsis spinimana* (Zoropsidae) has established a breeding population in the San Francisco Bay area.

Sparassids have tufted claws of specialized hairs at the tip of their legs, allowing them to cling to smooth surfaces. They run up walls and across ceilings. These spiders are highly vibration sensitive, quickly tracking any insect wandering into their path. Some large tropical species ambush tadpoles and small fish.

A Madagascar species, *Olios coenobita*, creates a home for itself in old snail shells. The shell is lifted off the ground and hung

Huntsman Spider, Australia
© Ryan Pike

in the air. One such shell was found to be 35 times the weight of its resident spider. In order to lift the shell, the spider methodically attaches webbing between the shell and a higher-placed branch. As the web dries, it contracts, raising the shell into the air. More webbing is attached, and higher the shell goes. The female spider lays her eggs in the shell, providing them with a safe retreat. When they are ready to move off on their own, they seek out their own appropriately sized shells. Each time they molt, they find a larger shell to inhabit.

Dancing White Lady Spider
© Thomas Neau

One desert sparassid exhibits remarkable homing ability. Found in the Namib desert, dancing white lady spiders (*Leucorchestris arenicola*) burrow into the desert sands. On calm dark nights, the large males set out in search of a meal or a mate. They prey on desert beetles and the like. Sand provides accurate vibratory detection, so prey can be determined up to three meters away. As it wanders, a male passes through other spiders' territories. It stops regularly to drum its legs and abdomen on the surface of the sand. This signals its presence to other males, as they all drum and maneuver around each other. These treks can reach 300 meters in a single night, but the spider always finds its way back to its home burrow. The spider navigates around obstacles and recognizes missed

turns. Scientists are studying this species in hopes of learning more about its navigational mechanism, which may offer insight for robotics technologies.

Another Namib desert sparassid is the wheel spider, *Carparachne aureoflava*. If the spider spots a predator while at the top of a sand dune, the spider tilts on its side, tucks in its legs, and rolls quickly down the sand. Wind assists gravity, allowing the spider to reach speeds up to a meter per second.

Green Huntsman Spider, India
© Maneesh Kaul

TWO-TAILED SPIDERS
HERSILIIDAE

The "two tails," or extra-long spinnerets for which these spiders are named, are not merely decorative. When approaching prey, perhaps entangled in a small patch of silk, the spider faces away from its dinner and crab-steps quickly around and around, all the while covering its meal in sticky threads from the spinnerets.

These spiders are also called tree-trunk spiders, because they are often sit camouflaged on bark or walls waiting in ambush. Species in this family are found in tropical and semi-tropical regions all around the world.

Two-Tailed Spider, Borneo
© Dr. Arthur Anker, Smithsonian Tropical Research Institute, Panama City, Panama

RECLUSES AND KIN
SICARIIDAE

Recluses are six-eyed spiders, with the eyes paired in three distinct clumps. They have fused chelicerae, with only their fangs being capable of movement. Prey is attacked and bitten to immobilize it. With larger prey, the spider backs away and waits for the venom to take effect before moving in to feed. Recluses prefer dry environments, and spin small mesh sheet-webs with a retreat. While these sheet-webs may trap wandering insects, recluses are primarily active stalkers.

The violin spiders (*Loxosceles* spp.) are infamous for their venomous bite. Most North American species are desert spiders with little human contact, but the brown recluse (*Loxosceles reclusa*) is known to inhabit homes. Too often, however, purported

Loxosceles rufescens
© Jorge Almeida

brown recluse bites are actually caused by other conditions. Cases of cutaneous anthrax, lyme disease, *Staphylococcus* infections, and bites from other species of spiders (*e.g.*, yellow sac spiders) have been misdiagnosed as brown recluse envenomations.

Many suspected bites are found far outside the known range of the brown recluse (southern central North America), so the supposed spider is suggested to have hitched a ride with transported cargo. Biologists have shown that this is an unlikely scenario. After studying homes infested with brown recluses within their native range, the scientists found that bites just did not occur under normal circumstances, even with several cases of direct human contact. The probability of a single vagrant recluse biting someone outside its native range is extremely low.

Similarly, spiders themselves are often misidentified as recluses. Arachnologists routinely receive spiders sent by the general public as "brown recluses," and identify them as any of several brownish wandering spiders (wolf spiders, sac spiders, huntsman spiders, spitting spiders, *etc.*). Irregular dark markings on the backs of such spiders are mistaken for the specific "violin" mark of a brown recluse.

You can determine that a spider is not a brown recluse if a) its body, not counting the legs, is larger than half an inch, b) there are visible hairs anywhere on the spider's body, c) it is found in a typical spider web, and d) there are visible markings other than a distinctive "violin" on the back of the carapace.

This is not to say, of course, that an actual bite from a recluse is insignificant. Recluse venoms contain necrotizing enzymes that destroy tissue and blood cells. Extent of damage may be aggravated by a patient's sensitivities to the bite. Fortunately, swift medical treatment will allay most difficulties, even halting ulceration of the skin. Fatalities from *Loxosceles* are rare, but other sicariids, such as the African crab-legged species, *Sicarius hahnii*, can be far more dangerous. That species, however, rarely interacts with humans. *Sicarius* species can be found in southern Africa and South America. During the day, *Sicarius* covers itself up with sand. It only coming out at night when its prey is active.

SPITTING SPIDERS
SCYTODIDAE

Spitting spiders are extraordinarily equipped for confronting prey and predators. Their cephalothorax houses glands for venom, silk production, and mucilage production. Depending on species, scytodids spin a web or search actively for prey. They have poor vision, so when they hunt, they raise their stilt-like forelegs, allowing sensitive hairs to detect prey movement. When they determine that prey is nearby, they approach to within an inch and spit a sticky net onto the quarry. This mucilaginous gum quickly dries and contracts, stopping the victim in its tracks. It was once thought this sticky net was itself poisonous, but recent studies suggest it is not always toxic, if at all. The spider alters the amount of spit depending on the size of the prey and how actively it resists capture. After ensuring its prey is immobilized, the spitting spider removes it from the spit, and swathes it in silk for feeding.

Scytodes thoracica, Germany
© Lorenz Kunze

Some species, as with *Scytodes pallida*, are also notable for their maternal care. The females carry their eggs in their chelicerae, even when dangerous predators are in the area. Direct care of the eggs greatly increases their chance for survival to hatching. This can be disadvantageous for the female, though, because they cannot spit to protect themselves when they carry the eggs. Chemical cues from known predators, however, induce the female spitting spider to stimulate, possibly through an enzymatic process, early hatching by the spiderlings. The female can then protect herself and her brood.

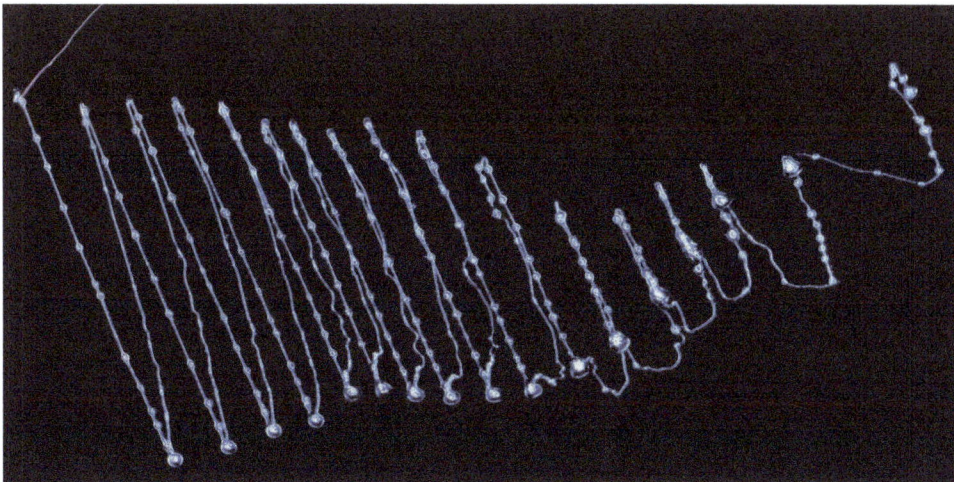

Scytodes Spit
© R. B. Suter, Vassar College

Beauty and Design in the World of Spiders

LITTLE WARRIORS
GNAPHOSIDAE, DYSDERIDAE, ZODARIIDAE

Small spiders often tackle prey close to their own size. Many, as with the ubiquitous ground spiders, or gnaphosids, commonly prey on other spiders. Gnaphosids are widely distributed and are commonly found at night hunting in leaf litter. Species often have specific habitat preferences, but as a family, they have adapted to a wide range of terrestrial ecosystems. Being small, gnaphosids are often prey for larger spiders. One species, *Herpyllus blackwalli*, uses its spinnerets to defend itself from the rear, shooting out a curtain of silk at an attacker. Another gnaphosid, *Drassodes lapidosus*, uses a similar technique to capture prey, making sure that its prey is well-entangled before getting close.

Zodariids show variety in their prey capture strategies. In North America, the genus *Lutica* (found in the coastal dunes of southern California, Baja, and the Channel Islands) are small ambush predators that create a sand-and-silk burrow. When small insects walk across the burrow, the spider rushes from the burrow to grab it, or will even lunge through the burrow's ceiling to grip the prey and drag it into the tube. *Zodarion rubidum*, an ant-specialist, actively hunts its prey. This species is an ant mimic, and practices a hit-and-run strategy. Because ants can be formidable adversaries for a small spider, it will bite the ant's rear leg and wait for its venom to paralyze the ant before moving in feed. European members of this species have even come up with a strategy to safely move through an ant colony while carrying its meal. When approached, the spider taps its front legs on the ant's antennae, then raises the captured prey so that the curious ant will note its scent. This multimodal deception allows the spider to retreat to a safe place and finish its meal.

The Australian zodariid *Habronestes bradleyi* specializes in feeding on the ant, *Iridomyrmex purpureus*. These ants are aggressive protectors of their colonies. During territorial fights with ants from other colonies, injured *Iridomyrmex* workers release alarm pheromones which attract more of their fellow colonists. *Habronestes* also detects these chemicals, using them to track down and capture the injured ant.

Dysderids are small spiders with large fangs. These are put to good use when small hard-shelled invertebrates are found. During the day, the spiders stay in their silk retreat on damp ground underneath stones, wood, or debris. At night, they actively forage for their preferred prey, isopods (pill or sow bugs).

Woodlouse Spider, *Dysdera crocata*
© Joseph Berger, ForestryImages.org

Zodarion styliferum
© Jorge Almeida

WANDERING SPIDERS
CTENIDAE

Ctenids can be found in most tropical regions. *Cupiennius* is one of the most intensively studied spider genera in the world, as its physiology and behavior are investigated as biological models. *Cupiennius* live on banana plants, *Agave*, and other monocots, and are finely attuned to receiving vibrations through the leaf surface. They use this to their advantage, not only locating prey, but sending vibratory courtship signals to each other. If a night-wandering male comes across a female's pheromone-rich dragline, he begins his quivering courtship. If she is in range of the vibrations, and willing to mate, she responds in kind, allowing the male to locate her in the tropical foliage.

Most ctenids are harmless, though they may cause painful bites. A serious exception, however, is the South American genus *Phoneutria*. These are aggressive spiders that have caused fatalities. They are sometimes called armed spiders, because upon being threatened, they display and wave their forelegs. Most cases of envenomation occur when the spider takes up residence where an unwary foot or hand may incite attack. As with most venomous species, children and the elderly are most at risk when bitten.

Phoneutria boliviensis, Ecuador
© Dr. Arthur Anker, Smithsonian Tropical
Research Institute, Panama City, Panama

Cupiennius getazi, Panama
© Dr. Arthur Anker, Smithsonian Tropical
Research Institute, Panama City, Panama

PIRATE SPIDERS
MIMETIDAE

Most pirate, or assassin, spiders specialize in preying on other spiders. They usually target web-building spiders, but will attack others. Mimetids have large spine-like hairs jutting in all directions off their long forelegs, allowing them to grab and cage their prey for a venomous bite.

When dusk sets, the pirate spider locates a suitable web, and waits on the outskirts. It may simply sit and wait until the resident spider steps past unawares. Or, it will imitate a struggling insect caught in the web. By plucking, tapping, pushing, and pulling the threads, the pirate spider draw the poor-sighted resident close. Sometimes, pirate spiders drive away a web-builder, take over its home, and feed on its egg-cases and captured prey.

Because web-weaving spiders rely on vibrations, they are attuned to even the slightest disturbance. Pirate spiders use this to their advantage to determine if a web has prey worth stalking. Mimetids create faint vibrations on the web by clutching the strands and sending out pulses, either by pushing up and down or by rotating the silk in a circle. These aren't strong enough to mimic trapped prey, but a resident spider will shift and move in place when the pulse is detected. This lets the pirate spider know that potential prey is available, and where it is located on the web.

The European pirate spider, *Ero furcata*, mimics more than thread-twitching prey. When it encounters a pair of courting theridiid spiders, it drives the male away and takes his place. The visually-impaired female does not realize that the arousing web vibrations no longer come from her romantic suitor. When she approaches close enough, the pirate spider lunges.

Mimetus sp.
© Kimberly R. Fleming

Mimetus sp. Feeding on Uloborid Eggsac
CC: Joe Lapp

Ero quadrituberculata
© Jorge Almeida

Cellar Spiders
Pholcidae

Rather than using orb-shaped webs in their prey-capture strategy, cobweb weavers spin sheet webs or erratic three-dimensional space webs to fill up open space. When found in the corners of a house, we call them cobwebs, but they are also discovered in shrubbery, tree branches, bark, rock outcroppings, and caves.

Pholcids are often called "daddy-long-legs" spiders, because they superficially resemble harvestmen (Opiliones). They are most commonly found in the corners of basements or attics, often living in small groups. These pholcid groups do not cooperate in prey capture, but expend less energy individually by sharing the responsibility of space web production. Hunger and size determine whether a pholcid will stay within a group or move on to build its own web. Smaller spiders that have difficulty competing with larger pholcids will abandon the web for their own hunting grounds. Big cellar spiders often move into large group webs, recognizing a better opportunity to compete for more prey.

Pholcid spider
© Vladimir Mucibabic

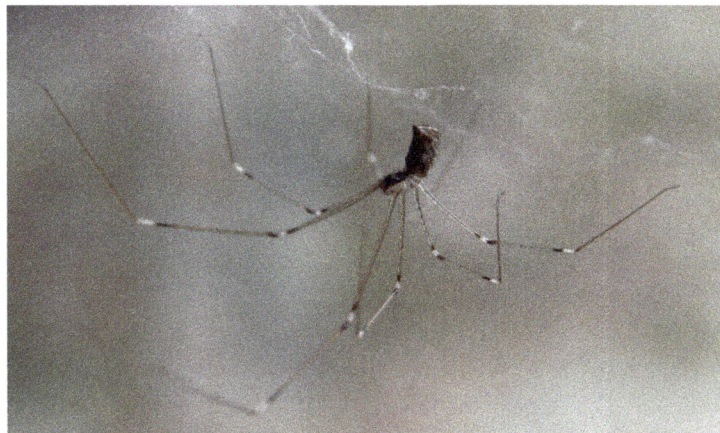

Pholcus phalangioides
© Jorge Almeida

In some species, where prey is not abundant and mates pair up on their own webs, male pholcids give up captured prey to the female. This "chivalrous" behavior provides greater nutrition to the egg-producing female. These males aggressively defend their mates from other potential sires.

While pholcids typically spin a curved sheet web for prey capture, females of the common cellar spider, *Holocnemus pluchei*, create a small globular dome web that they inhabit with their egg sac. This dome web surrounds them completely, aiding in the detection of potential predators. It also serves as a nursery and molting platform for the spiderlings after they hatch.

Pholcids often use physical motion in anti-predator tactics. Some twirl their bodies in circles, whirling their abdomen around in the air while holding on to strands of thread with their long legs. Others bounce by rapidly flexing and extending their legs, often changing the frequency and rate of bounce during longer sessions. These defense systems create a moving target and make it difficult for approaching predators to gain foothold on the silk. If this doesn't work, the pholcid just drops from the web and hides.

LACE-WEB WEAVERS
AMAUROBIIDAE, DESIDAE, DICTYNIDAE, OECOBIIDAE

Lace-web or mesh-web weavers are cribellate spinners, forming strands of wooly-combed silk around the primary catching threads. Insects that land on these threads become entangled as their feet catch in the wooly silk. These threads become clumps of bluish silk, though some are spun into a funnel-like form.

Lace-Web Weaver, *Amaurobius* sp.
CC: Rob Partington

Oecobiid spiders are very tiny, and are rarely noticed though they often live in houses. Most adult *Oecobius* species are less than 2mm in length. Oecobiids weave a small disk of silk on a wall that is used as cover until prey is detected.

One group of large-fanged desids, in the genus *Desis*, has found a home between the high and low tide marks on tropical beaches. During high tide, they shelter wherever they find air pockets among the submerged coral debris. When the tide is low, they hunt the tidal pools for small marine invertebrates like sea lice and shrimp. Another group of meshweavers with a specialized habitat are the dictynid spiders of the genus *Cicurina*. These are cave spiders, with many species endemic to particular cave systems.

The Amaurobiidae include several species of social and subsocial spiders. *Amaurobius socialis* live in colonies throughout its entire life. Subsocial spiders live cooperatively only for a period of time after hatching. One particular spider, *Amaurobius ferox*, shows parental care and juvenile coopera-tion. After the spiderlings emerge from their cocoon, they mass around the mother, stimulating her to laying a second batch of eggs. These eggs are immediately consumed by the spiderlings. When full, the spiderlings move off into the web. This feeding keep sibling aggression to a minimum. The mother provides yet another meal, however. After the spiderlings have spent several days digesting the eggs, they molt. A few days after molting, they become hungry again. The mother spider begins a sequence of interactions with her offspring, triggering them to attack and feed on her. This gives the spiderlings a nutritional boost that cannot be matched by their own attempts at capturing prey. It is only several days later that group cohesion fades, and the spiderling become more aggressive to each other. They move off, then, for lives on their own.

SHEET-WEB WEAVERS
LINYPHIIDAE

Linyphiids are small in size, spinning sheets and hammocks of silk that drape the grass or foliage. These may be flat or bowl-shaped, or a combination of both. These webs sometimes form distinctive shapes or patterns, lending the species names like hammock spider, bowl and doily spider, platform spider, and filmy dome spider.

Florinda coccinea, Black-tailed Red Sheetweaver
© Lew Scharpf

The sheet webs are composites of several layers of threading hung from an outer framework of silk scaffolding. The sheet webs are not sticky, but drops of glandular liquid are deposited during formation which evaporate to bond the sheets together. The spider hangs from underneath this silk surface, waiting for an insect to entangle itself. The spider strikes from below, pulling its prey through the silk and wrapping it for consumption.

In the bowl and doily spider, *Frontinella pyramitela*, juveniles and mature females create an intricate webwork of silk, forming a

Bowl and Doily Spider Web
© Lew Scharpf

Frontinella communis, Bowl and Doily Spider
© Photography by Marcy Kellar

bowl-shaped web overlying a flat sheet web. Above the bowl is an irregular vertically-rising barrier web. As insects hit the barrier web, they fall down into the bowl and are grabbed by the underhanging spider. The lowest sheet web helps protect the spider in case of predatory attack from below.

Unlike most web-weavers, where a male spider's priority is usually just finding a mate, male bowl and doily spiders actively feed after maturity, but rely on insects captured in female webs. They must compete with the female for prey. Often, they remain in the web after mating, even though they have no need to guard the female. In this species, the first male to mate with the female will be the father of her spiderlings. This isn't the case with many other spiders.

Female bowl and doily spider add a pheromone to her web strands that attracts males. When males locate an immature female, they guard her until her final molt. They may have to fight off other males that are attracted to the web. Once the female is reproductively mature, the male begins his courtship. A system ensues of waving legs, bobbing the abdomen, and plucking the silk threads. His actions communicate to the female that he isn't potential prey, provide an assessment of his worth as a mate, and stimulate her into acceptance.

The sierra dome spider, *Neriene litigiosa*, has a dome-shaped web with randomly spun capture threads above. When mature, females of this species may also add a pheromone to their webbing to attract potential mates. A male spider, however, that successfully locates a suitable female, strategically counteracts the attraction scent. He cuts most of the strands of the dome web, rolling the threads together into a few thick bundles. This dramatically reduces the evaporation of the female spider's web-based pheromone, making it difficult for other wandering males to locate and challenge the male who was first on the scene.

Males in other spider groups sometimes utilize gifts of wrapped prey to a female they are courting. With the linyphiid *Baryphyma pratense*, the male gives something of himself. The female grasps the male spider by the head, and sinks her fangs into his cephalic glands. She intakes some of the resulting hemolymph as they mate, but without any further harm to the male. He survives the encounter while apparently providing his mate with a nutritious meal to begin her egg development. Certain other linyphiids secrete attracting or arousing chemicals from their cephalic glands, which sometimes must be triggered by the female salivating on the male's head.

COMB-FOOTED SPIDERS
THERIDIIDAE

Theridiids are space-web weavers. Their irregular three-dimensional webbings are built underneath cover (rock, log, rafters, etc.) or have a hidden retreat nearby. In most species, these web-works are supported structurally with radiating strands, and threaded throughout with tensely-held silk lines that are often viscid, or sticky. Some space webs are large complex structures, while others are very small, or just a few lines. Strand webs are commonly spun on branches, targeting small arthropods that wander on from the tree. In one South American theridiid, *Phoroncidia studo*, a single line is stretched out, with only part of it coated with a sticky substance. This sticky silk holds a chemical attractant, making this single thread a very effective flytrap.

Steatoda fulva locates a mound of harvester ants, building its web next to an entrance while the ants are resting. When the ants emerge, a few get stuck and release a chemical alarm. Soldier ants have trouble freeing the victims, so they retreat and close up that entrance. When the activity dies down, *Steatoda* emerges from hiding and wraps up the captured ants.

Most theridiids (about 2200 species known) are solitary by nature, and only convene for courtship and mating. A few species are semi-social, show maternal care, or tolerate each other for periods of time if prey is abundant. Even fewer, like some *Anelosimus* species, are communal, living in colonies and attacking prey in small groups.

The common house spider, *Achaearanea tepidariorum*, spins webs in corners, attics, basements, and window-sills. It captures a wide range of species, and will feed on invertebrates six times its own length. Should prey be dangerous (as with spider-eating assassin bugs), it will be carefully cut out of the web and dropped to the ground.

Steatoda paykulliana, False Widow Spider, Israel
© Shai Pilosof

Social Spider, *Anelosimus* sp.
CC: Joe Lapp

Enoplognatha ovata, Candy Stripe Spider
© Doxa / Shutterstock

Widow Spiders
Latrodectus species

Widows are an infamous theridiid group of about thirty venomous species. Found throughout the world, these often colorful spiders go by many names: button spiders, katipo spiders, red-back spiders, brown widows, black widows, and red widows. Five species of widows are known from North America: the western, northern, and southern black widows, Florida's endemic red widow, and the cosmopolitan brown widow.

These spiders were given the name "widows," because of a recognized proclivity in some species for a female to cannibalize the male spider after mating. Of course, aggressive female behavior is present in most spiders, and males must be very careful not to trigger an attack. Most times, a healthy male widow will escape before the female turns on him, and there is some justification for noting that early studies of widow spiders had a built-in bias—researchers did not allow the male to leave the female's territory after mating, giving her ample time to capture and eat him.

In the red-back widow, *Latrodectus hasselti*, there is good reason to suspect that males allow, or intentionally trigger, female cannibalization. This

Latrodectus hesperus, Western Black Widow
© Jeff Cleveland

sacrifice increases the male's chances for paternity, provides nutrients for potential offspring, and reduces competition by other males. Males of certain other spider genera offer silk-wrapped prey as a tasty gift to their potential mate; these male red-backs have the option of offering themselves. Unlike other widows, male red-backs are usually functionally sterile after the first mating encounter, so they have more at stake with this reproductive opportunity. Because red-back females have dual sperm-storage organs, a single mating would only result in 50% paternity for a successful male red-

Latrodectus hasselti, Red-Back Spider
© Renate Micallef

back. Researchers discovered that a red-back male will often constrict ("pinch in") the top of his abdomen during the first mating, making it more difficult for the female's fangs to kill him right away. Then, he uses the second of his paired copulatory organs to mate again, so that all of the offspring has a greater chance of being his own.

Before they can mate, male black widows must first find a suitable female. As with many cobweb-building spiders, female black widows add a pheromone to their threads after maturity. Each species uses and blends varying chemical attractants, and males are capable of distinguishing females from different populations of the same species. Once the male encounters this pheromone, the chemicals induce courtship behavior. He begins vibrating his abdomen while snipping strands of the female's web and balling up the threads. The rhythmic shaking lulls the female into passive mode, allowing him to get close, slip a few strands of his own thread onto her (commonly called a "bridal veil"), and mate. By dropping from the web after mating, the male is able to escape. Because he has changed the layout of her webbing, the female has less chance of relocating him for a later meal.

Widow venoms are neurotoxic. *Latrodectus mactans* venom contains multiple neurotoxins, each specifically targeting different organisms. This species feeds on many different arthropods and even on smaller vertebrates. Alpha-latrotoxin is the neurotoxin that targets vertebrates. It binds to presynaptic proteins in the prey's nervous system, triggering excessive release of acetylcholine and other secreted neurotransmitters. There is a briefly delayed pain reaction, but this quickly intensifies throughout the body. In human envenomations, symptoms can include vomiting, coordination loss, spasms and cramping, respiratory distress, and muscular rigidity. While bites can be fatal, prompt treatment greatly helps a successful recovery.

A Male Widow Spider, South Carolina
© Kimberly R. Fleming

Latrodectus mactans, Southern Black Widow
© Terry L. McCormick

Latrodectus hasselti, and Captured Skink
© Joel Bohm

Kleptoparasitic Theridiids
Argyrodes species

While there are other spiders that live as kleptoparasites, none are as commonly encountered as *Argyrodes*. These spiders live on the outskirts of another species' orb-web or cob-web, often attaching their own small web to that of the resident spider's web. They feed on leftover scraps or captured insects that are too small for the resident spider's liking (commensalism), or steal prey that the resident spider wants for itself (kleptoparasitism). Sometimes, *Argyrodes* will kill and eat the resident spider itself, remaining in the web to feast on any further insect prey. At least some species of *Argyrodes* will eat silk from their host spider's web during times when few insects are captured. Larger webs attract

A Small *Argyrodes* Waiting for a Meal
© EcoView

Argyrodes sp., Dewdrop Spider
© Jorge Almeida

several *Argyrodes*, and if too many show up, the resident spider may abandon its web and seek another location for its web-spinning.

Large orb-webs can be dangerous places to stalk, as other predators target orb-weaving spiders, so *Argyrodes* are often camouflaged. Some have thin elongated bodies resembling twigs. Others have round pointed abdomens that are silver in coloration. These are called dewdrop spiders, and possibly mimic drops of rain on the webbing.

Species of *Argyrodes* are often host-spider specific for a given habitat and season. This specificity may be the result of complicating factors with certain web types: sticky threads have to be circumvented, and vibration transmission quality may be poor on the outskirts of some webs.

A few species of *Argyrodes* do spin simple webs for capturing other spiders. *Argyrodes colubrinus* lays out a few long threads in hopes that ballooning spiders will use them as a launch-point, giving *Argyrodes* time to sneak up and grab them. *Argyrodes attenuatus* lays out similar non-sticky lines, attracting both ballooning spiders and small "trapeze" flies which use the silk as a resting place. The spider slowly creeps up, then readies a strand of sticky silk which it uses to grab and hold its quarry.

With most *Argyrodes*, mating is fairly simple and lacks elaborate courtship. This is probably due to females lacking the aggressive predatory tendencies that most other spiders have. Males actively pursue females, so they will chase and grapple with potential rivals or disrupt other males during courting. This fighting rarely escalates to injury or fatalities, as *Argyrodes* doesn't play "winner takes all." "Losing" males will sneak back later and court the female while the "winner" is busy elsewhere.

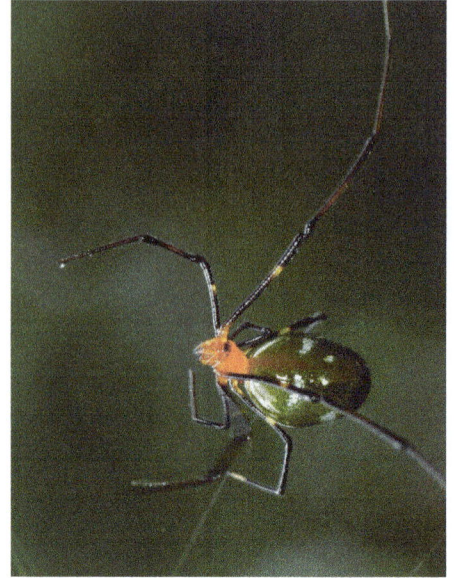

Argyrodes cf. *miniaceus,*
Singapore
© Yixiong Cai

Happy Face Spiders
Theridion grallator in Hawaii

This species of spider is found only on four islands in Hawaii. It offers biologists opportunity to investigate genetic polymorphism, as at least eleven different color and pattern morphs appear within the different island populations. The typical morph is the standard "yellow" spider without any markings other than some black spotting or blotches. These make up about 70% of any of the island's *T. grallator* populations.

The other 30% may have red blotches, marks, or rings in various positions on their abdomen, additional black markings, blobs, or rings, white abdominal blotches, or black lines on the carapace. In some variations, sometimes with more than one morph present, the spider's abdomen markings have the appearance of a "happy face," which gives them their name.

Genetically, some morphs are dominant, others co-dominant with each other. The typical yellow morph is recessive to all the other varieties. On the island of Maui, studies indicate that most of the color morphs are the result of alleles at a single color locus, with the characters being expressed evenly throughout the population. On the island of Hawaii, a few morphs are sex-linked, expressing themselves in either male or female spiders. This points to two unlinked autosomal loci being responsible for the color variation.

Theridion grallator
© Jack Jeffrey Photography

ORB-WEAVING SPIDERS

Orb-weavers usually have poor eyesight, but are adept at detecting web-borne and air-borne vibrations. They can, for example, recognize a predatory wasp's wingbeats as it flies toward them. Orb webs provide an effective method of targeting and capturing prey despite visual limitations. Many orb-weavers, like the black-and-yellow garden spider, *Argiope aurantia*, hang upside down in the center of their web waiting for entangled prey to vibrate the strands. This causes the spider to spin and face the target. Other orb-weavers sit in a retreat attached to the orb with signal threads, charging to the orb when prey bounces on the web.

Once the presence of prey is known, the orb-weaving spider plucks the strands in front of it, to stabilize them and determine whether the prey is in that sector. It moves forward until it is close enough to recognize the prey. Common prey species elicit specific predatory behavior: grasshoppers are quickly swathed in blankets of silk and envenomated, butterflies or moths may be bitten first before their loose-scaled wings get free, while smaller prey may not require wrapping. The prey is usually brought to the hub to be eaten, though it may be stored in the outer web for later feeding.

Most orb-weavers are sexually dimorphic. The females are much larger and are the primary orb-weavers. Males may not even have a similar shape. After maturity, the small males begin searching for mates as most have only a few months left before they die. They locate the web of a suitable female, and take residence on the outskirts, waiting for the right opportunity for courtship. Because females will sometimes attack approaching suitors, the male spider often waits until the female is feeding, is torpid from molting, or has been habituated to the male's presence. Male orb-weavers in a few species, as in *Nephiles clavipes*, will sometimes offer a nuptial gift to the female they are courting. A small prey item, tightly wrapped, is provided to occupy the female's attention while the male mates.

Orb-weavers do exhibit a level of intelligence. Memory of captured prey has been determined in araneids and tetragnathids. One case of parasitic deterrence by the Mexican colonial spider, *Metepeira incrassata*, illustrates anticipatory behavior. A parasitic fly wishing to lay its egg on the

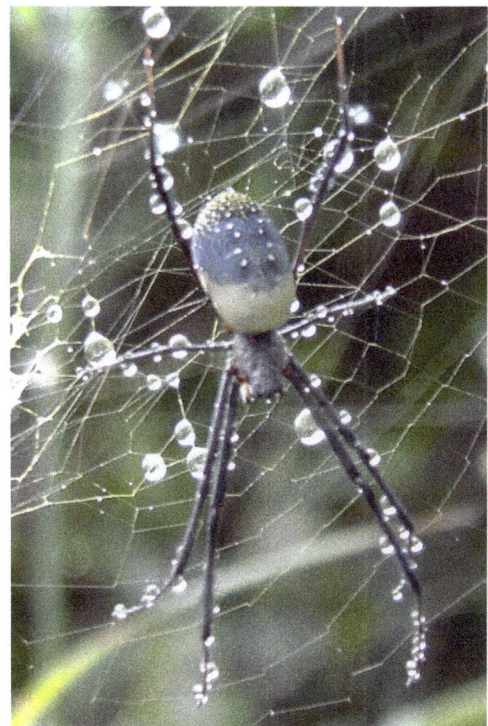

Underside of an Orb-Weaver, Brazil
© Andrea Paccini

Orb-Weaver, South Africa
© Humphrey Grimmett

spider's egg-sac lands on the hub or signal-line of the orb web, vibrating it to draw the female away from its retreat. The spider's immediate response is genetically programmed, so it rushes from the retreat to the center of the orb looking for captured prey. The fly tries to distract the spider long enough for it to deposit the egg. After prolonged attack cycles, spiders recognize the fly's attack technique, and respond by cutting the orb's signal-line. This allows them to guard the egg-sac in peace until the parasitic fly has left the area.

While orb-weavers are capable of learning from past experience, their ability to spin orb webs derives from their genes. The webs of different species can often be differentiated visually. This genetic foundation for web-spinning is exploited by one parasitic wasp, of the ichneumonid genus *Hymeno-epimecis*. This wasp targets the orb-weaver *Plesiometa argyra*, first paralyzing it, and then laying an egg on its abdomen. The paralysis soon wears off, and the spider goes about its business, continuing to build orb webs for the next couple weeks. During this period, the egg hatches and the wasp larva feeds on the spider's blood. On the night the larva is ready to molt, it chemically modifies the spider's web-building behavior. Instead of creating a typical orb-web, the spider constructs sturdy and less-elaborate webbing on a horizontal plane. If the web isn't built correctly with the strongest silk threads, it won't last long during tropical rainstorms. When the web is finished, the larva molts, then kills and eats its spider host. When sated, the larva spins its own cocoon, hanging it from the sturdy webbing.

Structurally, the orb web's outermost threads are the framework for the webbing. At strategic points, anchor threads fasten framing threads together, attaching them to vegetation or other solid objects. Radii are threads that cross from the frame to the middle, or hub, of the web. After the radii are in place, the spider adds spiraling silk to the orb. While there is often an initial non-sticky spiral, to be eaten away later, the primary capture spiral is sticky. In most cases, the spiral is not

Tetragnathid Capturing Dragonfly
CC: Joe Lapp

Orb Web
© Martina Berg

Argiope bruennichi, European Wasp Spider
© Sven Peter

Phonognatha graeffei (Tetragnathidae)
CC: Steve Baty

as densely formed around the hub of the web. The orientation of the orb depends on the species. Most araneids spin vertical orbs, while tetragnathid orbs are often at an angle. Uloborids usually place orbs on the horizontal plane. The spiral capture thread is responsible for keeping prey trapped until the spider reaches it. The radii, while important structurally, are necessary to let the spider know that prey is available. The vibrations of capture threads do not trigger a feeding attack by resident spiders. Only radii movement elicits a response.

The spider often lets out a single line of free-floating silk. This is used when the spider needs to cross gaps where the orb has been damaged. For species which do not rest at the hub of the orb, signal threads extend from the orb to a retreat. Barrier webs are found in some species, irregular threading to the sides of the orb that lacks sticky silk. These provide greater stability for the orb and are sometimes used to store wrapped prey.

While most orb webs are circular, the webs of a few genera are highly modified into vertical ladder-like arrangements. The web's hub is often at the top of a long, thin, tapering sheet that contains both radii and crosshatching spiral threads. In *Scoloderus cordatus*, the ladder-web is inverted, in that the hub is at the bottom of the sheet. This sheet can be several feet in length and only inches wide. This configuration is particularly effective for moths. They struggle to free themselves when they hit the top of the web, sliding down the threads, all the while losing wing scales that could otherwise keep them from sticking to the threads. Eventually, their descent is halted as they are held fast to the silk.

The adhesive strands in an orb web must be renewed daily. At the end of the predatory cycle (day or night, depending on whether the spider is diurnal or nocturnal), most species take down their orb. Much of this is eaten, allowing the web's proteins and amino acids to be recycled for later use. At the beginning of the next cycle, the spider builds a new orb. In colonial orb-weavers, only the sticky orb is renewed; framework and interconnecting strands are maintained over long periods.

One common element in an orb-weaver's web is the stabilimenta, or additional silk structures woven into a web. These come in different configurations: circular disks, spirals, and zig-zags are common forms. In some species, as with *Cyclosa* orb-weavers (or "trash" spiders), insect remains and other debris are added to the web. Naturalists suggest a multitude of explanations for these patterns. In some cases, stabilimenta strengthen the web structurally, provide a retreat behind which the spider can cool down, keep a predator from recognizing the spider as potential prey, or draw prey to the spider. Some patterns are more important at different times in a spider's life, within different habitats, and under different environmental conditions.

Cruciform stabilimenta consists of four wedges of zig-zag lines branching out like the spokes of a pinwheel. Sometimes these connect in the middle, other times they are separate. These mimic large pairs of legs, creating the impression that a spider hanging from the middle of the orb's hub is much bigger than it really is. This helps deter predators who might consider the meal biting off more than they can chew.

Top Left, Clockwise: Cross Stabilimenta, Singapore (© Phil Date); Shield Stabilimenta (© Wendy Pastorius); Circular Stabilimenta, Virgin Islands (© Craig Smith); Zig Zag Stabilimenta (© Grégory Smellinckx).

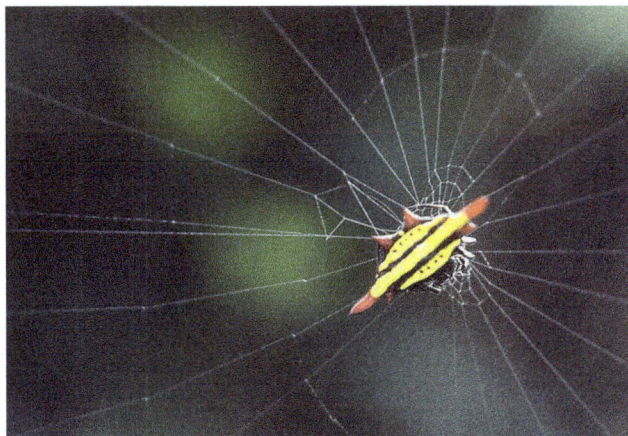

Top Left, Clockwise: *Gasteracantha* sp. (© Frank Hatcher); *Micrathena clypeata*, Red Crab Orbweaver, Ecuador (© Dr. Arthur Anker, Smithsonian Tropical Research Institute, Panama City, Panama); *Micrathena* sp., Ecuador (© Dr. Arthur Anker, STRI); *Gasteracantha* sp., China (© Cedric Basset); *Gasteracantha cancriformis* (© Paul Wolf)

Beauty and Design in the World of Spiders

TYPICAL ORB-WEAVERS
ARANEIDAE

Araneids are the most commonly recognized orb-weavers, though not all make orbs. Large colorful *Argiope* species can be encountered in fields and forest edges throughout North America. *Araneus* species are often found making orb webs in barns and older architecture. While most species are solitary predators that wait for prey to become entangled, some araneids break the rules by congregating in colonies or by actively enticing their prey.

Orb-weavers require locations with abundant prey. If a location proves barren, the spider moves to a new spot. Many species favor mixed vegetation, particularly ecotonal habitats like the edges of forests or along the sides of a road. Complex vegetation structures allow different life stages to find suitable prey and environmental conditions. The orientation of the web is based on the environment, as well. It is situated to maximize sunlight in cool shade, and to minimize the sun's rays in bright open areas. Damaging winds must also be taken into consideration, affecting the web's inclination from the vertical.

Agalenatea redii
© David Acosta Allely

Prey species captured are usually a consequence of location and orb size. When a specific prey species is abundant, spiders modify the mesh size and capture area of the orb, or adjust the height from the ground, to selectively catch that species. Butterflies, moths, dragonflies, grasshoppers, flies, mosquitoes, beetles, bees, and true bugs are fair game. Rarely, the silk entangles small vertebrates. If a distasteful or dangerous creature is captured, the spider clips off the strands holding the victim, allowing it to escape.

The North American asterisk-web weaver, *Wixia ectypa*, creates an extreme example of a reduced orb web. It consists of a few strands of radii with a small circular hub. The radii are each fixed along small branches. None of the silk strands are sticky. Instead, they carry vibrations to the waiting spider from small insects walking over the threads among the foliage. Upon being alerted, *Wixia* rushes toward its prey and begins swathing silk as it circles around and around the branch. Its victim finds itself stuck tight and unable to defend itself.

Another araneid, the Tasmanian *Paraplectanoides crassipes*, also targets pedestrian invertebrates, but spins a stranger web. After attaching some radii and a small hub horizontally near the ground, the spider expands the webbing into a three-dimensional orb-like structure, incorporating leaves and other debris into the latticework. As roaches and other small insects crawl into the web-ball's only opening, the spider inside snatches them.

Orb-weaver
© Pawel Pietras

Orb-weaver, Underside
© Aaron Bailey

The ability of stabilimenta to draw prey to a web is especially interesting. Because stabilimenta reflects UV light, it is effective in a forest setting, where insects mistake the reflectance for a patch of sunlight. With some insects, particularly bees, the UV reflectance may trigger flower pattern recognition.

Just as the role of stabilimenta continues to be debated, so does the coloration of visually conspicuous spiders. Many orb spiders are brightly colored, like the yellow-and-black *Argiope aurantia*, or *Gasteracantha* spiny orb-weavers. Just because the coloration is eye-catching to us, however, does not mean that it will warn off other species. Patterns of colors may resemble floral clusters, attracting small pollinators. Bees use achromatic vision at long distances, switching to chromatic vision at close range. One Asian tetragnathid orb-weaver, *Nephila pilipes*, takes advantage of this. While its brightly contrasting colors might be detected far off, a bee's compound eye blurs forms at that distance. When the bee is closer, switching to chromatic vision, the spider's dark spots blend into the ambient background, leaving a collection of yellow spots and stripes that are unlikely to be recognized as a spider. Rather, those patterns are similar enough to floral patterns to attract bees into the web.

Of course, not all araneids are colorful. One Australian species of the genus *Carepalxis* shows ingenious mimicry, resembling the Eucalyptus tree's gumnut. The nocturnal spider rests at the edge of a twig during the day, with its round abdomen pressed close and legs tucked in. Its smooth brown abdomen includes a dark round pigmentation spot, resembling a nut's opening after its seeds have been released.

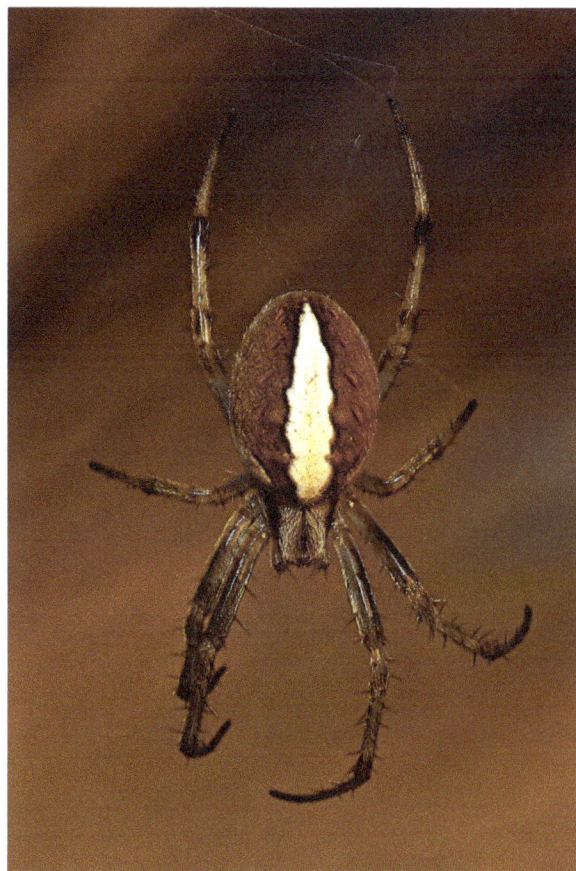

Eriophora sp., Waikoloa, Hawaii
© Bill Adams

Micrathena sagittata
© Bill Beatty

One Mexican orb-weaver, *Metapeira incrassata,* has been studied extensively due to its coloniality. While some of these spiders will be found in their own individual territories, this species often is found in groupings of individually-maintained orb webs. Colony size is related to environmental conditions and prey abundance, but may reach thousands of spiders. Tolerance of close neighbors may seem unusual in a competitive predator, but these colonies do have some advantages. While small prey is easily trapped in an orb web, larger prey (especially with higher flight velocity) may break through a single web. In clumped web groupings, breaking through would only lead to further entanglement, and the prey's kinetic energy would diffuse as it bounces around before being caught for good in the web. This is called the "ricochet effect." Another advantage is an early warning system from predators. If a predatory wasp strikes the colony, spiders under immediate attack jump from their webs, creating vibrations throughout the interconnected orb web system that are recognized as warning signals.

Beauty and Design in the World of Spiders

Bolas Spiders

Not all araneids weave orb webs. Adult female spiders in the New World genera *Mastophora* and *Agastostichus*, the African genera *Cladomelea* and *Acantharachne*, and the Australian genus *Ordgarius* have a remarkable prey-catching technique. They target their prey with a long spun thread attached to a sticky ball of webbing.

Cladomelea akermani
© John Roff, Hilton College, South Africa

The nocturnal female *Mastophora* is the best studied of these genera. It searches for a suitable location under a leaf or twig where it spins a single-strand horizontal platform from which to hang. It then spins a long thread that hangs from this platform. At the end of this hanging web strand is attached a globulous mass of adhesive web. When this is ready, the spider begins to emit chemical cues from its body that mimic the sexual pheromones of certain moths. As it hangs from the platform, the female waits with the thread of this "bolas" in its grip. When male moths come close, the spider swings the thread at them, hoping to catch the sticky ball on their wings. If successful, the spider descends the thread and feeds, or wraps up the moth and stores it on the platform thread while it continues hunting.

These sticky balls of *Mastophora* are formed by liquid of several viscosities surrounding the core of balled-up webbing. A lower viscosity liquid penetrates the scales of a moth's wings, while a higher viscosity liquid sticks the thread to the scales. The internal configuration of the webbing is responsible for this particular thread's elasticity. The silk expands like an extended spring, rather than stretching like a rubber band. If the bolas spider misses a target or hits itself accidentally, the liquid is carefully ingested for recycling.

Juvenile bolas spiders also emit a chemical component, but one that differs from the adult females' mimicry. Small immature *Mastophora* are not able to spin an effective bolas at their size, so they concentrate on a different prey, flies. The spiderlings hang over the edge of a leaf with their front legs held motionless in front of them. As flies are attracted to the mimicking chemistry, they are grabbed when they come close enough. Juveniles of different species of *Mastophora* attract different species of flies, with psychodid flies being a primary target. As juvenile females grow larger, they switch tactics and chemical mimicry to target male moths. Adult male *Mastophora* are much smaller than adult females, and they retain their ability to attract flies. This allows them to feed after maturity, until the slower-maturing females are ready for courtship. The entire process from cocoon emergence to maturity, reproduction, and death happens over the course of a single year.

These are not the only araneids to use pheromone-like chemicals to attract prey. The genus *Kaira* also creates these chemical attractants, though they don't make a bola. Instead, they make a small temporary web from which they hang, forming a small basket with its extended legs. When a male moth flies into the basket, it is snatched, wrapped, and stored between the spider's legs until the spider has collected enough for the night.

Two Strange Orb-Weavers

Poltys sp., Tree-Stump Spider, India
© Bir Bahadur and Maneesh Kaul

Arachnura sp., Scorpion-Tailed Spider,
Australia
© Matthew Dallinger

Beauty and Design in the World of Spiders

LONG-JAWED ORBWEAVERS
TETRAGNATHIDAE

The Tetragnathidae includes the long-jawed orb-weavers and orchard spiders, typically identified by their long chelicerae. Many have long narrow bodies, while others look like the typical round orb-weavers. Twig-like mimicry is not uncommon among the former. Spiders in the genus *Tetragnatha* are probably the most widely distributed and abundant orb-weavers around the globe. Some are habitat specialists. *Tetragnatha elongata*, for example, is a riparian specialist that can only survive in close vicinity to water.

Most are solitary hunters, but *Nephila clavipes*, the golden silk spider, builds interconnected web groupings. The coloration of the silk, especially in densely packed web groupings, may help steer passing animals away from the strands.

The giant Asian wood spider, *Nephila maculata*, is one of a few orb-weaving spiders commonly eaten by native peoples. Roasted, salted, and eaten whole (or sometimes just the abdomens), the spiders are said to have a nutty flavor.

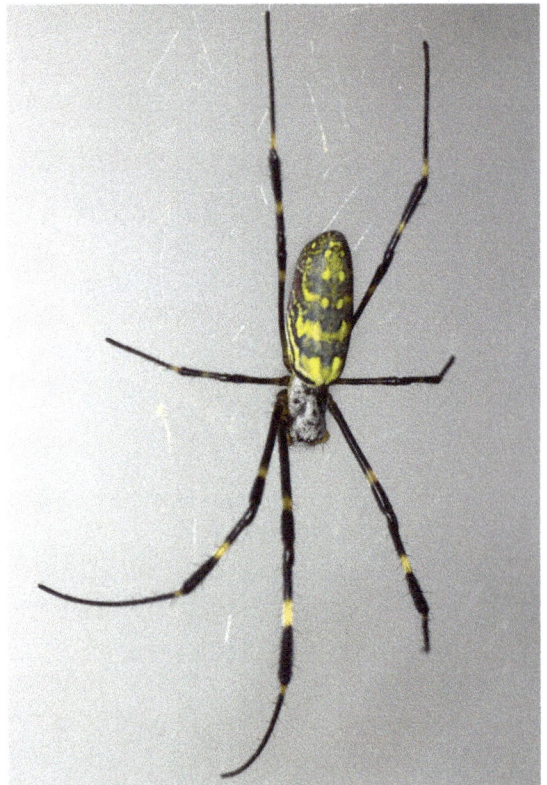

Top Right, Clockwise:
Tetragnatha extensa, United Kingdom
© Paul Whillock

Nephila clavata, Japan
© Sean Barley

Leucauge venusta, Orchard Spider
© Terri Heisele

HACKLED ORBWEAVERS
ULOBORIDAE

The Uloboridae are the only complete family of spiders lacking venom glands. Capture behavior of many species compensates for this by increasing the time spent wrapping prey. Their long forelegs help to accomplish this task. Typical uloborids include the feather-legged orb-weaver (*Uloborus glomosus*), triangle-web spiders (*Hyptiotes* spp.), and twig-like orb-weavers (*Miagrammopes* spp.).

Uloborids are sometimes called hackled orb-weavers because of their silk's special characteristic. The sticky silk of other orb-weavers is induced by droplets of a moist adhesive on the silk itself. Not so with uloborid sticky strands. Uloborids are the only orb-weavers to spin cribellar webbing. While the spider's posterior spinnerets lay out a pair of core axial threads, these are surrounded by thinner threads from the median spinnerets, and then completely covered with thousands of tiny microfibrils from the cribellar spigots as the spider draws its setal comb over the cribellum. This technique sometimes gives a bluish tint to the silk.

Uloborids that build simple webs have stickier strands than those that build more complex orbs because they add more microfibrils to each individual thread. These microfibrils are responsible for the adhesiveness, though the mechanism is poorly understood. Electrostatic charge is one possibility.

One African species of *Miagrammopes* sets out a single thread of sticky cribellar silk, then sits at one end pulling the silk taut. When an insect stops to rest on the silk, the spider lets go of the thread. The tension is released, and the thread springs back and forth, catching up the hapless victim in the sticky microfibrils.

Feather-Legged Orb-weaver and Web
© Lew Scharpf

Beauty and Design in the World of Spiders

OGRE-FACED SPIDERS
DEINOPIDAE

Deinopids spin a different type of web for their nocturnal hunting. Rather than create a large orb web and wait for insects to fly into it, an ogre-faced spider spins a small flexible net that it holds with its first two pairs of legs. As it hangs from a strand of thread just above the vegetation, it waits for a passing insect. If prey comes walking along the leaves, the spider quickly reaches forward, wrapping it up. If the insect comes flying, the spider moves just as quickly. These spiders are finely attuned to aerial vibrations, striking backward and casting the net around the victim. This net stretches up to ten times its original size.

The genus *Deinopis* also has what may be the largest simple eyes of any land invertebrate, a pair of posterior median eyes providing acute nocturnal vision. The optical sensitivity comes from a highly sensitive photoreceptor membrane, which forms in the eyes at the beginning of every night and deteriorates at sunrise.

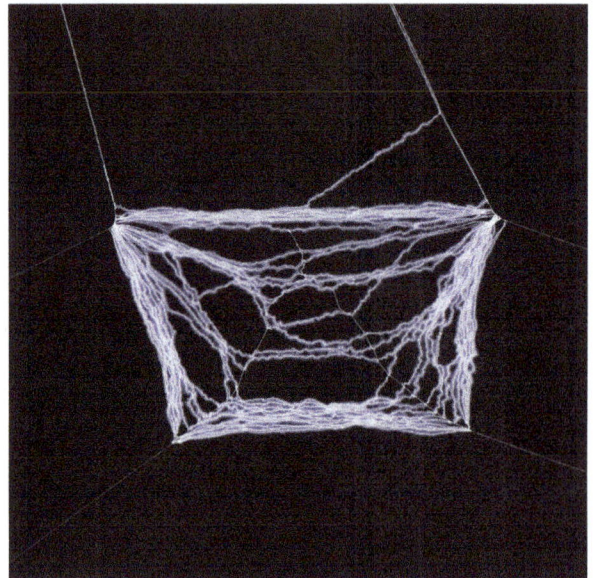

Deinopis sp., including a net-casting adult, a bug's-eye view of a *Deinopis* net, and a female forming an egg-sac.
All images, © Gary Warner, Australia

STUDYING SPIDERS

Spiders provide ample opportunity for students to study natural history and practice scientific methods. Observation and experimentation of spider behavior, surveys of local species, and investigation of web building will enhance biology education. The best place to start is to look for spiders in their natural habitat and watch them, taking notes. Students can record, sketch, and photograph where they are found, what they look like, what they are doing, and how they interact with other species.

Advanced students should consider joining a spider survey project, if one is available in their state or region. This would involve collecting specimens and recording accurate collection data. While some spiders can be visually identified, other species will require microscopic examination by experienced spider-hunters. Still, spiders can usually be identified to family level using diagnostic keys. Spider survey projects should provide identification materials and instructions for preserving specimens. Of course, collecting on public land often requires a permit. Collecting on private lands should be done with written permission of the land-owners. Spider enthusiasts should be aware of any protected or threatened species, taking care not to collect them. Trapdoor spiders, for example, are particularly vulnerable to over-collection.

Micrathena sagittata
© Lew Scharpf

Students of all ages can collect spider webs. Flat orb-webs can be found along the edges of forests and fields. First, gently blow or tap the resident spider until it falls to the ground and scurries off into the brush. A light coating of hairspray strengthens the strands. Talcum powder or flour can be dusted onto the threads. Then, carefully press a large sheet of dark construction paper flat against the web. This secures the main orb, and excess strands can be cut off. Later, the page can be covered in clear contact paper.

Some artists collect orb webs for permanent display. They first coat the orb with metallic spray paint. Before it dries, this is carefully pressed against a large pane of lightly clear-varnished glass. After the web is collected, the back of the glass is painted black, and the project is framed.

Webs of different shapes can be collected through photographs. This is best done in the early morning, when dew on the silk threads has not yet evaporated. You can find dew-covered webs in the grass, among the shrubbery, on tree bark, and in flower gardens.

Students enthralled with spiders will, with appropriate permission, find a pet tarantula is a rewarding educational experience. Several species are suitable for the attentive beginning keeper. The rosehair tarantula (*Grammostola rosea*), the pinktoe tarantula (*Avicularia avicularia*), the Costa Rican zebra tarantula (*Aphonopelma seemani*), and the curlyhair tarantula (*Brachypelma albopilosum*) are all good starter arachnids. More aggressive species should be left to advanced hobbyists. A healthy young captive-bred tarantula can be kept in a naturalistic vivarium, and observed as it grows. Tarantulas are not throw-away pets. Be prepared to keep it for many years. A large tank with a tightly clasped lid will provide plenty of room for most tarantulas. Before buying one, be sure to research the environmental conditions and necessary supplies for the species you want. There are several good books and websites available on tarantula husbandry.

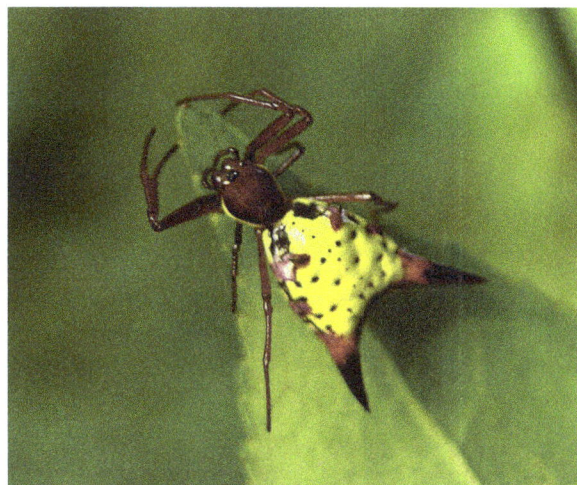

BEAUTY AND DESIGN

Spiders survive and thrive in a world filled with predators, prey, and competing species. Physical and behavioral variability allows these creatures to colonize vastly diverse habitats, while selective pressures differentiate populations by winnowing out the least advantageous traits for that particular environment.

This is a straightforward, if simplistic, assessment of the facts. But, facts alone are not enough for an engaging examination of the natural world and its history. Facts need to be interpreted, which is done by placing them within the context of a philosophy or worldview for evaluation, discussion, and debate. Two polarized worldviews currently prominent are philosophical naturalism, which asserts that only what is material is real and that everything can be explained by strictly natural processes and laws, and creationism, which contends that there is a God who created and sustains physical reality and reveals Himself to us through nature (generally) and the Bible (specifically). There are many reasons why the origins debate will not end anytime soon, not least of which is that there is an wide range of perspectives between those two beliefs, within the philosophical gradient accompanying science. One key point of contention, however, stems from the evolutionary charge that no religious belief can be considered scientific, while non-Darwinians argue that naturalism, itself, is religious.

This short book is not the best place to examine the philosophical underpinnings of the origins debate, or to illustrate the often-overlooked history of religious support for scientific endeavors, but I think it is appropriate to note that the perspective of someone who recognizes the existence of purpose, beauty, and design in nature, even in the be-

Platorid Crab Spider, Ecuador
© Morley Read

Green Lynx Spider
© Lew Scharpf

havior and morphology of small creatures like spiders, can only enhance the value of scientific study. Dr. E. O. Wilson, the famed sociobiologist and humanist, recently wrote a book trying to persuade religious people of the value of conservation. His intentions were good, but by continuing to maintain a false dichotomy between religion and science (when in fact, the dichotomy is between theism and naturalism), he undercuts his effectiveness. This problem is not uncommon, with philosophical naturalists often stating that religion itself is a barrier to scientific inquiry. Christian apologist Dr.

Francis Schaeffer noted the error in that argument: "To say that God communicates truly does not mean that God communicates exhaustively. ... Thus, as far as our position in the universe is concerned, though the infinite God has said true things concerning the whole of what He has made, our knowledge is not thereby meant to be static. Created in His image, we are rational and logical and, as such, we are able to and intended to explore and discover further truth concerning creation."

So, from a theistic perspective, we are to share God's delight in creation's diversity and ingenuity. Creation also provides us with endless natural designs to stimulate ideas for solving real world problems. Spider silk and spider venoms offer a wealth of prospective discoveries for manufacturing, pharmacology, and other technologies. Just as important, we have a responsibility as stewards of the Earth and its inhabitants, which calls for us to maintain healthy ecological systems. To do that, we must learn how those systems work, and understand the importance of every species.

For the non-theist, design and beauty in nature, of course, are byproducts of the human imagination, particularly in view of a natural world filled with death and suffering. Robert Frost advanced this in his poem, *Design* (1932), describing his encounter with a white spider holding a white moth on a white flower. Was it coincidence or malevolent design that brought a moth to its death? He ends the poem:

What had that flower to do with being white,
The wayside blue and innocent heal-all?
What brought the kindred spider to that height,
Then steered the white moth thither in the night?
What but design of darkness to appall?—
If design govern in a thing so small.

John Bunyan, in the 1600s, also addressed natural evil—even focusing on the spider as an example. In his poem, *The Sinner and the Spider* (1686), a man and spider debate the origins and purpose of natural evil, the effects of man's folly on himself and nature, and what man can learn from creation. The spider proclaims:

Thus in my ways God wisdom doth conceal,
And by my ways that wisdom doth reveal.

And so, the man acknowledges one purpose for design in nature:

Thou art my monitor, I am a fool;
They learn may, that to spiders go to school.

Woodlouse Spider Attacking a Black Widow
© Jamie Voetsch

Argiope
© Lew Scharpf

Beauty and Design in the World of Spiders

BIBLIOGRAPHY

Aiken, Marie, and Frederick A. Coyle. 2000. Habitat distribution, life history and behavior of *Tetragnatha* spider species in the Great Smoky Mountains National Park. *Journal of Arachnology* 28: 97-106.

Airamé, Satie, and Petra Sierwald. 2000. Hunting and feeding behavior of one *Heteropoda* species in lowland rainforest on Borneo (Araneae, Sparassidae). *J. Arachnology* 28: 251-253.

Allan, Rachel A., Mark A. Elgar, and Robert J. Capon. 1996. Exploitation of an ant chemical alarm signal by the zodariid spider *Habronestes bradleyi* Walckenaer. *Proceedings of the Royal Society of London B* 263: 69-73.

Amalin, Divina M., Jonathan Reiskind, Jorge E. Peña, and Robert McSorley. 2001. Predatory behavior of three species of sac spiders attacking citrus leafminer. *J. Arachnology* 29: 72-81.

Andrade, Maydianne C. B. 1996. Sexual selection for male sacrifice in the Australian redback spider. *Science* 271: 70-72.

Andrade, Maydianne C. B., and Erin M. Banta. 2002. Value of male remating and functional sterility in redback spiders. *Animal Behaviour* 63: 857-870.

Andrade, Maydianne C. B., Lei Gu, and Jeffrey A. Stoltz. 2005. Novel male trait prolongs survival in suicidal matings. *Biology Letters* 1: 276-279.

Arango, Angélica M., and Victor Rico-Gray. 2000. Population structure, seasonality, and habitat use by the green lynx spider *Peucetia viridans* (Oxyopidae) inhabiting *Cnidoscolus aconitifolius* (Euporbiaceae). *J. Arachnology* 28: 185-194.

Austin, A. D. 1984. Life history of *Clubiona robusta* L. Koch and related species (Araneae, Clubionidae) in south Australia. *J. Arachnology* 12: 87-104.

Ayala, Francisco J. 2000. Darwin and the teleology of nature. Haught, John F., ed. *Science and Religion in Search of Cosmic Purpose*. Washington, D. C.: Georgetown University Press.

Baird, Craig R., and Robert L. Stolz. 2002. Range expansion of the hobo spider, *Tegenaria agrestis*, in the northwestern United States (Araneae, Agelenidae). *J. Arachnology* 30: 201-204.

Barham, James. 2001-2005. Natural selection and teleology. *ISCID Encyclopedia of Science and Philosophy*. www.iscid.org/encyclopedia/

Becker, Nathan, et al. 2003. Molecular nanosprings in spider capture-silk threads. *Nature Materials* 21: 278-283.

Benjamin, Suresh P., and Samuel Zschokke. 2002. Untangling the tangle-web: web construction behavior of the comb-footed spider *Steatoda triangulosa* and comments on phylogenetic implications (Araneae: Theridiidae). *Journal of Insect Behavior* 15(6): 791-809.

Benjamin, Suresh P., Marcel Düggelin, and Samuel Zschokke. 2002. Fine structure of sheet-webs of *Linyphia triangularis* (Clerck) and *Microlinyphia pusilla* (Sundevall), with remarks on the presence of viscid silk. *Acta Zoologica* 83: 49-59.

Bennett, Robert G. 1987. Systematics and natural history of *Wadotes* (Araneae, Agelenidae). *J. Arachnology* 15: 91-128.

Bergman, Jerry. 2007. Lack of fossil evidence for arthropod evolution is a major difficulty for Neo-Darwinism. *Creation Research Society Quarterly* 43(4): 222-230.

Bergman, Jerry. 2001. Evolutionary naturalism: an ancient idea. *TJ* 15(2): 77-80.

Binford, G. J. 2001. An analysis of geographic and intersexual chemical variation in venoms of the spider *Tegenaria agrestis* (Agelenidae). *Toxicon* 39: 955-968.

Binford, Greta J., and Michael A. Wells. 2003. The phylogenetic distribution of sphingomyelinase D activity in venoms of Haplogyne spiders. *Comparative Biochemistry and Physiology Part B* 135: 25-33.

Blanchong, Julie A., et al. 1995. Chivalry in pholcid spiders revisited. *J. Arachnology* 23: 165-170.

Bleckmann, Horst, and Thomas Lotz. 1987. The vertebrate-catching behaviour of the fishing spider *Dolomedes triton* (Araneae, Pisauridae). *Animal Behaviour* 35: 641-651.

Blest, A. D. 1987. The copulation of a linyphiid spider, *Baryphyma pratense*: does a female receive a blood meal from her mate? *Journal of Zoology, London* 213: 189-192.

Bond, Jason E., and Frederick A. Coyle. 1995. Observations on the natural history of an *Ummidia* trapdoor spider from Costa Rica (Araneae, Ctenizidae). *J. Arachnology* 23: 157-164.

Bond, Jason E., and Brent D. Opell. 2002. Phylogeny and taxonomy of the genera of south-western North American Euctenizinae trapdoor spiders and their relatives (Araneae: Mygalomorphae, Cyrtaucheniidae). *Zoological Journal of the Linnean Society* 136: 487-534.

Bonthrone, Karen M., et al. 1992. The elasticity of spiders' webs is due to water-induced mobility at a molecular level. *Proc. Royal Soc. London B* 248: 141-144.

Brach, Vincent. 1976. Subsocial behavior in the funnel-web wolf spider *Sosippus floridanus* (Araneae: Lycosidae). *The Florida Entomologist* 59(3): 225-229.

Breene, Robert Gale, III. 1997. *Recluse Spiders and the Hobo Spider in North America*. Artesia, NM: American Tarantula Society.

Breene, Robert Gale, III. -2005. *Respiration*. American Tarantula Society. Atshq.org.

Breene, Robert G., and Merrill H. Sweet. 1985. Evidence of insemination of multiple females by the male black widow spider, *Latrodectus mactans* (Araneae, Theridiidae). *J. Arachnology* 13: 331-335.

Bucaretchi, Fábio, et al. 2000. A clinico-epidemiological study of bites by spiders of the genus *Phoneutria*. *Rev. Inst. Med. trop. S. Paulo* 42(1): 17-21.

Buchli, Harro H. R. 1969. Hunting behavior in the Ctenizidae. *American Zoologist* 9: 175-193.

Cangialosi, Karen R. 1990. Life cycle and behavior of the kleptoparasitic spider, *Argyrodes ululans* (Araneae, Theridiidae). *J. Arachnology* 18: 347-358.

Clark, David L., and George W. Uetz. 1993. Signal efficacy and the evolution of male dimorphism in the jumping spider, *Maevia inclemens*. *Proceedings of the National Academy of Sciences* 90(24): 11954-11957.

Clark, Robert J., Robert R. Jackson, and Bruce Cutler. 2000. Chemical cues from ants influence predatory behavior in *Habrocestum pulex*, an ant-eating jumping spider (Araneae, Salticidae). *J. Arachnology* 28: 309-318.

Clements, Reuben, and Daiqin Li. 2005. Regulation and non-toxicity of the spit from the pale spitting spider *Scytodes pallida* (Araneae: Scytodidae). *Ethology* 111(3): 311-321.

Comstock, John Henry. 1912. *The Spider Book*. Garden City, NJ: Doubleday, Page, and Co.

Costa, Fernando G., and Fernando Pérez-Miles. 1992. Notes on mating and reproductive success of *Ceropelma longisternalis* (Araneae, Theraphosidae) in captivity. *J. Arachnology* 20: 129-133.

Costa, Fernando G., and Fernando Pérez-Miles. 1998. Behavior, life cycle and webs of *Mecicobothrium thorelli* (Araneae, Mygalomorphae, Mecicobothriidae). *J. Arachnology* 26: 317-329.

Coyle, Frederick A. 1986. The role of silk in prey capture by nonaraneomorph spiders. in William A. Shear, ed. *Spiders: Webs, Behavior, and Evolution*. Stanford, CA: Stanford University Press.

Coyle, Frederick A., and Wendell R. Icenogle. 1994. Natural history of the Californian trapdoor spider genus *Aliatypus* (Araneae, Antrodiaetidae). *J. Arachnology* 22: 225-255.

Coyle, Frederick A., and William A. Shear. 1981. Observations on the natural history of *Sphodros abboti* and *Sphodros rufipes* (Araneae, Atypidae), with evidence for a contact sex pheromone. *J. Arachnology* 9: 317-326.

Coyle, Frederick A., Theresa C. O'Shields, and Daniel G. Perlmutter. 1991. Observations on the behavior of the kleptoparasitic spider, *Mysmenopsis furtiva* (Araneae, Mysmenidae). *J. Arachnology* 19: 62-66.

Craig, Catherine L. 1997. Evolution of arthropod silks. *Annual Review of Entomology* 42: 231-267.

Cramer, Ken. 2005. *Identifying recluses*. Brown Recluse Spider Project. Monmouth College. http://department.monm.edu/biology/recluse-project/identify.htm

Cushing, Paula E. 1997. Myrmecomorphy and myrmecophily in spiders: a review. *The Florida Entomologist* 80(2): 165-193.

Cushing, Paula E., and Richard G. Santangelo. 2002. Notes on the natural history and hunting behavior of an ant eating zodariid spider (Arachnida, Araneae) in Colorado. *J. Arachnology* 30: 618-621.

Cutler, Bruce. 2003. A new subspecies of *Philodromus rufus* (Araneae, Philodromidae). *J. Arachnology* 31: 142-144.

Eberhard, William G. 1979. *Argyrodes attenuatus* (Theridiidae): a web that is not a snare. *Psyche* 86: 407-413.

Eberhard, William G. 1980. The natural history and behavior of the bolas spider *Mastophora dizzy-deani* sp. n. (Araneidae). *Psyche* 87(3-4): 143-169.

Eberhard, William G. 1981. The single line web of *Phoroncidia studo* Levi (Araneae: Theridiidae): a prey attractant? *J. Arachnology* 9: 229-232.

Eberhard, William G. 1987. How spiders initiate airborne lines. *J. Arachnology* 15: 1-9.

Eberhard, William G. 2000. Spider manipulation by a wasp larva. *Nature* 406: 255-256.

Edwards, G. B. 2001. The present status and a review of the brown recluse and related spiders, *Loxosceles* spp. (Araneae: Sicariidae), in Florida. Florida Department of Agriculture and Consumer Services. Division of Plant Industry. *Entomology Circular* No. 406.

Edwards, G. B., and R. R. Jackson. 1993. Use of prey-specific predatory behavior by North American jumping spiders (Araneae, Salticidae) of the genus *Phidippus*. *J. Zoology* 229: 709-716.

Edwards, G. B., J. F. Carroll, and W. H. Whitcomb. 1974. *Stoidis aurata* (Araneae: Salticidae), a spider predator of ants. *The Florida Entomologist* 57(4): 337-346.

Edwards, Robert L., and Eric H. Edwards. 1990. Observations on the natural history of a New England population of *Sphodros niger* (Araneae, Atypidae). *J. Arachnology* 18: 29-34.

Elgar, Mark A., and Rachel A. Allan. 2004. Predatory spider mimics acquire colony-specific cuticular hydrocarbons from their ant model prey. *Naturwissenschaften* 91(3): 143-147.

Eskov, K. Yu., and S. L. Zonstein. 2000. The first ctenizoid mygalomorph spiders from Eocene Baltic amber (Araneida: Mygalomorphae: Ctenizidae). *Paleontological Journal* 34(Supp. 3): S268-S274.

Eubanks, Micky D., and Gary L. Miller. 1992. Life cycle and habitat preference of the facultatively arboreal wolf spider, *Gladicosa pulchra* (Araneae, Lycosidae). *J. Arachnology* 20: 157-164.

Fernández-Montraveta, Carmen, and Mariano Cuadrado. 2003. Cohabitation between an adult male and a subadult female in a burrowing wolf spider (Araneae, Lycosidae). *J. Arachnology* 31: 135-138.

Fink, Linda S. 1984. Venom spitting by the green lynx spider, *Peucetia viridans* (Araneae, Oxyopidae). *J. Arachnology* 12: 372-373.

Firstman, Bruce. 1973. The relationship of the chelicerate arterial system to the evolution of the endosternite. *J. Arachnology* 1: 1-54.

Foelix, Rainer F. 1996. *Biology of Spiders*. 2nd edition. New York: Oxford University Press.

Foradori, Matthew J. 2002. Relation between the outer cover of the egg case of *Argiope aurantia* (Araneae: Araneidae) and the emergence of its spiderlings. *Journal of Morphology* 252(2): 218-226.

Forster, Lyn. 1982. Vision and prey-catching strategies in jumping spiders. *American Scientist* 70: 165-175.

Fowler, Harold G. 1984. Note on a clubionid spider associated with Attine ants. *J. Arachnology* 12(1): 117-118.

Furey, Robert E., and Susan E. Riechert. 1989. *Agelena consociata* (Araneae, Agelenidae) and its nest associates: insect cleaners. *J. Arachnology* 17: 240-242.

Furneaux, W. 1911. *Life in Ponds and Streams*. London: Longmans, Green, and Co.

Garb, Jessica E., Alda González, and Rosemary G. Gillespie. 2004. The black widow spider genus *Latrodectus* (Araneae: Theridiidae): phylogeny, biogeography, and invasion history. *Molecular Phylogenetics and Evolution* 31: 1127-1142.

Getty, Richard M., and Frederick A. Coyle. 1996. Observations on prey capture and anti-predator behaviors of ogre-faced spiders (*Deinopis*) in southern Costa Rica (Araneae, Deinopidae). *J. Arachnology* 24: 93-100.

Gilbert, Cole, and Linda S. Rayor. 1985. Predatory behavior of spitting spiders (Araneae: Scytodidae) and the evolution of prey wrapping. *J. Arachnology* 13: 231-241.

Gillespie, Rosemary G., and Geoffrey S. Oxford. 1998. Selection on the color polymorphism in Hawaiian happy-face spiders: evidence from genetic structure and temporal fluctuations. *Evolution* 52(3): 775-783.

Gonzaga, Marcelo de Oliveira, Adalberto José dos Santos, and Guilherme Fraga Dutra. 1998. Web invasion and araneophagy in *Peucetia tranquillini* (Araneae, Oxyopidae). *J. Arachnology* 26: 249-250.

Gorb, Stanislav N., and Friedrich G. Barth. 1994. Locomotor behavior during prey-capture of a fishing spider, *Dolomedes plantarius* (Araneae: Araneideae): galloping and stopping. *J. Arachnology* 22: 89-93.

Gorder, Pam Frost. 2003. Scientists believe ancient arachnids may have spun silk like modern spiders. Press release. *Ohio State University Research News*.

Greene, Erick, Larry J. Orsak, and Doughlas W. Whitman. 1987. A tephritid fly mimics the territorial displays of its jumping spider predators. *Science* 236: 310-312.

Gregory, Jr., Ben M. 1989. Field observations of *Gasteracantha cancriformis* (Araneae, Araneidae) in a Florida mangrove stand. *J. Arachnology* 17: 119-120.

Gosline, J. M., et al. 1999. The mechanical design of spider silks: from fibroin sequence to mechanical function. *The Journal of Experimental Biology* 202: 3295-3303.

Griswold, Charles E. 1998. The nest and male of the trap-door spider *Poecilomigas basilleupi* (Araneae, Mygalomorphae, Migidae). *J. Arachnology* 26: 142-148.

Griswold, Charles E., et al. 1999. Towards a phylogeny of entelegyne spiders (Araneae, Araneomorphae, Entelegynae). *J. Arachnology* 27: 53-63.

Griswold, Charles E., and Darrell Ubick. 2001. Zoropsidae: a spider family newly introduced to the USA (Araneae, Entelegynae, Lycosoidea). *J. Arachnology* 29: 111-113.

Guinea, G. V., et al. 2005. Stretching of supercontracted fibers: a link between spinning and the variability of spider silk. *J. Experimental Biology* 208: 25-30.

Harland, Duane P., and Robert R. Jackson. 2000. 'Eight-legged cats' and how they see—a review of recent research on jumping spiders (Araneae: Salticidae) *Cimbebasia* 16: 231-240.

Harland, Duane P., and Robert R. Jackson. 2002. Influence of cues from the anterior medial eyes of virtual prey on *Portia fimbriata*, an araneophagic jumping spider. *J. Experimental Biology* 205: 1861-1868.

Hartmeyer, Siegfried. 1998. Carnivory in *Byblis liniflora* revisited II: the phenomenon of symbiosis on insect trapping plants. *Carnivorous Plant Newsletter* (ICPS) 27(4): 110-113.

Hauber, Mark E. 2002. Conspicuous colouration attracts prey to a stationary predator. *Ecological Entomology* 27: 686-691.

Hawthorn, Anya C., and Brent D. Opell. 2003. van der Waals and hygroscopic forces of adhesion generated by spider capture threads. *J. Exp. Biology* 206: 3905-3911.

Hebets, Eileen A., and George W. Uetz. 2000. Leg ornamentation and the efficacy of courtship display in four species of wolf spider (Araneae: Lycosidae). *Behavioral Ecology and Sociobiology* 47: 280-286.

Heiling, Astrid M., and Marie E. Herberstein. 2004. Floral quality signals lure pollinators and their predators. *Annales Zoologici Fennici* 41: 421-428.

Heiling, Astrid M., Marie E. Herberstein, and Lars Chittka. 2003. Crab-spiders manipulate flower signals. *Nature* 421: 334.

Henschel, Joh R. 2002. Long-distance wandering and mating by the dancing white lady spider (*Leucorchestris arenicola*) (Araneae, Sparassidae) across Namib dunes. *J. Arachnology* 30: 321-330.

Herberstein, M. E., et al. 2000. The functional significance of silk decorations of orb-web spiders: a critical review of the empirical evidence. *Biological Review* 75: 649-669.

Hieber, Craig S., R. Stimson Wilcox, Jay Boyle, and George W. Uetz. 2002. The spider and fly revisited: ploy-counterploy behavior in a unique predator-prey system. *Behavioral Ecology and Sociobiology* 53: 51-60.

Hodge, Maggie. 1984. Anti-predator behavior of *Achaearanea tepidariorum* (Theridiidae) towards *Stenolemus lanipes* (Reduviidae): preliminary observations. *J. Arachnology* 12: 369-370.

Holdsworth, Andrew R., and Douglass H. Morse. 2000. Mate guarding and aggression by the crab spider *Misumena vatia* in relation to female reproductive status and sex ratio. *American Midland Naturalist* 143: 201-211.

Hölldobler, Bert. 1970. *Steatoda fulva* (Theridiidae), a spider that feeds on harvester ants. *Psyche* 77: 202-208.

Jackson, Robert R. 1978. The life history of *Phidippus johnsoni* (Araneae: Salticidae). *J. Arachnology* 6: 1-29.

Jackson, Robert R. 1998. Spider-eating spiders. *American Scientist* 86: 350-357.

Jackson, Robert R. & Mary E. A. Whitehouse. 1986. The biology of New Zealand and Queensland pirate spiders (Araneae, Mimetidae): aggressive mimicry, araneophagy and prey specialization. *J. Zoology, London* (A), 210: 279-303.

Jackson, R. R., et al. 1993. Anti-predator defences of a web-building spider, *Holocnemus pluchei* (Araneae, Pholcidae). *J. Zoology, London* 229: 347-352.

Jaeger, P. 2001. A new species of *Heteropoda* (Araneae, Sparassidae, Heteropodinae) from Laos, the largest huntsman spider? *Zoosystema* 23(3): 461-465.

Jakob, E. M. 2004. Individual decisions and group dynamics: why pholcid spiders join and leave groups. *Animal Behaviour* 68: 9-20.

Johnson, Stephen R. 1996. Use of coleopteran prey by *Phidippus audax* (Araneae, Salticidae) in tall-grass prairie wetlands. *J. Arachnology* 24: 39-42.

Kaston, B. J. 1965. Some little known aspects of spider behavior. *American Midland Naturalist* 73(2): 336-354.

Kasumovic, Michael M., and Maydianne C. B. Andrade. 2004. Discrimination of airborne pheromones by mate-searching male western black widow spiders (*Latrodectus hesperus*): species- and population-specific responses. *Canadian Journal of Zoology* 82: 1027-1034.

Killebrew, Don W., and Neil B. Ford. 1985. Reproductive tactics and female body size in the green lynx spider, *Peucetia viridans* (Araneae, Oxyopidae). *J. Arachnology* 13: 375-382.

Kim, Kil Won, and Chantal Roland. 2000. Trophic egg laying in the spider, *Amaurobius ferox*: mother-offspring interactions and functional value. *Behavioral Processes* 50: 31-42.

Kim, Kil Won, Chantal Roland, and André Horel. 2000. Functional value of matriphagy in the spider *Amaurobius ferox*. *Ethology* 106: 729-742.

Klook, Carl T. 2001. Diet and insectivory in the "araneophagic" spider, *Mimetus notius* (Araneae: Mimetidae). *American Midland Naturalist* 146: 424-428.

Knight, David P., and Fritz Vollrath. 2002. Spinning an elastic ribbon of spider silk. *Philosophical Transactions of the Royal Society of London B* 357: 219-227.

Kovoor, J., and L. Zylberberg. 1980. Fine structural aspects of silk secretion in a spider (*Araneus diadematus*). I. Elaboration in the pyriform glands. *Tissue Cell.* 12(3): 547-556.

Levi, Herbert W. 1993. The orb-weaver genus *Kaira* (Araneae: Araneidae). *J. Arachnology* 21: 209-225.

Li, Daiqin. 2002. Hatching responses of subsocial spitting spiders to predation risk. *Proc. Royal Society London B* 269: 2155-2161.

Lim, Matthew L. M., and Daiqin Li. 2006. Extreme ultraviolet sexual dimorphism in jumping spiders (Araneae: Salticidae). *Biol. J. Linn. Soc.* 89(3): 397-406.

Linskens, H. F., F. Ciampolini, and M. Cresti. 1993. Spider webs as pollen traps. *Proceedings of the Koninklijke Nederlandse Akademie van Wetenschappen* 96(4): 415-425.

Locht, A., M. Yáñez, and I. Vázquez. 1999. Distribution and natural history of Mexican species of *Brachypelma* and *Brachypelmides* (Theraphosidae, Theraphosinae) with morphological evidence for their synonymy. *J. Arachnology* 27: 196-200.

Lockley, Timothy C., Orrey P. Young, and Jane Leslie Hayes. 1989. Nocturnal predation by *Misumena vatia* (Araneae, Thomisidae). *J. Arachnology* 17: 249-251.

Maddison, Wayne P., and Gail E. Stratton. 1988. Sound production and associated morphology in male jumping spiders of the *Habronattus agilis* species group (Araneae, Salticidae). *J. Arachnology* 16: 199-211.

Main, Barbara York. 1999. Notes on the biogeography and natural history of the orbweaving spider *Carepalxis* (Araneae, Araneidae), including a gumnut mimic from southwestern Australia. *J. Arachnology* 27: 183-188.

Marshall, Samuel D., and George W. Uetz. 1990. Incorporation of urticating hairs into silk: a novel defense mechanism in two neotropical tarantulas (Araneae, Theraphosidae). *J. Arachnology* 18: 143-149.

Mather, Monica H., and Bernard D. Roitberg. 1987. A sheep in wolf's clothing: tephritid flies mimic spider predators. *Science* 236: 308-310.

Moore, Clovis W. 1977. The life cycle, habitat and variation in selected web parameters in the spider, *Nephila clavipes* Koch (Araneidae). *American Midland Naturalist* 98(1): 95-108.

Morewood, Wm. Dean, Kelli Hoover, and James C. Sellmer. 2003. Predation by *Achaearanea tepidariorum* (Araneae: Theridiidae) on *Anoplophora glabripennis* (Coleoptera: Cerambycidae). *The Great Lakes Entomologist* 36(1-2): 31-34.

Morse, Douglass H. 1981. Prey capture by the crab spider *Misumena vatia* (Clerck) (Thomisidae) on three common native flowers. *American Midland Naturalist* 105(2): 358-367.

Morse, Douglass H. 1983. Foraging patterns and time budgets of the crab spiders *Xysticus emertoni* Keyserling and *Misumena vatia* (Clerck) (Araneae: Thomisidae) on flowers. *J. Arachnology* 11: 87-94.

Morse, Douglass H. 1985. Nests and nest-site selection of the crab spider *Misumena vatia* (Araneae, Thomisidae) on milkweed. *J. Arachnology* 13: 383-390.

Morse, Douglass H. 1988. Interactions between the crab spider *Misumena vatia* (Clerck) (Araneae) and its ichneumonid egg predator *Trychosis cyperia* Townes (Hymenoptera). *J. Arachnology* 16: 132-135.

Morse, Douglass H. 1995. Changes in biomass of penultimate-instar crab spiders *Misumena vatia* (Araneae, Thomisidae) hunting on flowers late in the summer. *J. Arachnology* 23: 85-90.

Morse, Douglass H. 2001. Harvestmen as commensals of crab spiders. *J. Arachnology* 29: 273-275.

Nelson, John M., and Ramon L. Lake. 1976. Observations on the crab spiders (Araneae: Thomisidae) at the Lake Bixhoma Reservoir and adjacent Tulsa County. *Proceedings of the Oklahoma Academy of Sciences* 56: 56-58.

Ono, Hirotsugu. 1999. Spiders of the genus *Heptathela* (Araneae, Liphistiidae) from Vietnam, with notes on their natural history. *J. Arachnology* 27: 37-43.

Opell, Brent D. 1994. Factors governing the stickiness of cribellar prey capture threads in the spider family Uloboridae. *J. Morphology* 221: 111-119.

Opell, Brent D. 2002. Estimating the stickiness of individual adhesive capture threads in spider orb webs. *J. Arachnology* 30: 494-502.

Opell, Brent D., Jamel S. Sandidge, and Jason E. Bond. 2000. Exploring functional associations between spider cribella and calamistra. *J. Arachnology* 28: 43-48.

Ortega-Escobar, J., and A. Muñoz-Cuevas. 1999. Anterior median eyes of *Lycosa tarentula* (Araneae, Lycosidae) detect polarized light: behavioral experiments and electro-retinographic analysis. *J. Arachnology* 27: 663-671.

Oxford, G. S., and R. G. Gillespie. 1996. Genetics of colour polymorphism in *Theridion grallator* (Araneae: Theridiidae), the Hawaiian happy-face spider, from Greater Maui. *Heredity* 76(1996): 238-248.

Oxford, G. S., and R. G. Gillespie. 1996. The effects of genetic background on the island-specific control of a colour polymorphism in *Theridion grallator* (Araneae: Theridiidae), the Hawaiian happy-face spider. *Heredity* 76(1996): 257-266.

Oxford, G. S., and R. G. Gillespie. 1998. Evolution and ecology of spider coloration. *Annu. Rev. Entomol.* 43: 619-643.

Oxford, Geoff S., and Rosemary G. Gillespie. 2001. Portraits of evolution: studies of coloration in Hawaiian spiders. *BioScience* 51(7): 521-528.

Palmer, Jacqueline M. 1985. The silk and silk production system of the funnel-web mygalomorph spider *Euagrus* (Araneae: Dipluridae). *J. Morphology* 186: 195-207.

Palmer, Jacqueline M., Frederick A. Coyle, and Frederick W. Harrison. 1982. Structure and cyto-chemistry of the silk glands of the mygalomorph spider *Antrodiaetus unicolor. J. Morphology* 174: 269-274.

Peakall, D. B. 1971. Conservation of web proteins in the spider, *Araneus diadematus. Journal of Experimental Zoology* 176: 257-264.

Pekár, Stano, and Jirí Král. 2002. Mimicry complex in two central European zodariid spiders (Araneae: Zodariidae): how *Zodarion* deceives ants. *Biological Journal of the Linnean Society* 75(4): 517-532.

Penney, David, and Paul A. Selden. 2002. The oldest linyphiid spider, in lower Cretaceous Lebanese amber (Araneae, Linyphiidae, Linyphiinae). *J. Arachnology* 30: 487-493.

Penney, David, C. Philip Wheater, and Paul A. Seldon. 2003. Resistance of spiders to Cretaceous-Tertiary extinction events. *Evolution* 57(11): 2599-2607.

Pickford, Martin. 2000. Fossil spider's webs from the Namib Desert and the antiquity of *Seothyra* (Araneae, Eresidae). *Ann. Paléontol.* 86(3): 147-155.

Pollard, Simon D. 1984. Egg guarding by *Clubiona cambridgei* (Araneae, Clubionidae) against conspecific predators. *J. Arachnology* 11: 323-326.

Pollard, Simon D., Mike W. Beck, and Gary N. Dodson. 1995. Why do male crab spiders drink nectar? *Animal Behaviour* 49: 1443-1448.

Prentice, Thomas R. 1992. A new species of North American tarantula, *Aphonopelma paloma* (Araneae, Mygalomorphae, Theraphosidae). *J. Arachnology* 20: 189-199.

Qin, Xiao-Xia, Kathryn J. Coyne, and J. Herbert Waite. 1997. Tough tendons: mussel byssus has collagen with silk-like domains. *The Journal of Biological Chemistry* 272(51): 32623-32627.

Rajashekhar, K. P., and K. P. Siju. 2003. Pretending to be a predator: wasp-like mimicry by a salticid spider. *Current Science* 85(8): 1124-1125.

Ramirez, Martin G. 1995. Natural history of the spider genus *Lutica* (Araneae, Zodariidae). *J. Arachnology* 23: 111-117.

Ramirez, Martin G., Estelle A. Wall, and Monica Medina. 2003. Web orientation of the banded garden spider *Argiope triasciata* (Araneae, Araneidae) in a California coastal population. *J. Arachnology* 31: 405-411.

Ramos, Margarita, Duncan J. Irschick, and Terry E. Christenson. 2004. Overcoming an evolutionary conflict: removal of a reproductive organ greatly increases locomotor performance. *Proc. National Academy Sciences* 101(14): 4883-4887.

Reiskind, Jonathan. 1969. Stereotyped burying behavior in *Sicarius. American Zoologist* 9: 195-200.

Reiskind, Jonathan. 1976. *Orsima formica*: A Bornean salticid mimicking an insect in reverse. *Bulletin of the British Arachological Society* 3(8): 235-236.

Riechert, Susan E., Rosemarie Roeloffs, and Arthur C. Echternacht. 1986. The ecology of the cooperative spider *Agelena consociata* in Equatorial Africa (Araneae, Agelenidae). *J. Arachnology* 14: 175-191.

Robinson, Michael H., and Barbara Robinson. 1976. The ecology and behavior of *Nephila maculata*: a supplement. *Smithsonian Contributions to Zoology* No. 218.

Robinson, Michael H., and Carlos E. Valerio. 1977. Attacks on large or heavily defended prey by tropical salticid spiders. *Psyche* 84(1): 1-10.

Rodriguez, Rafael L., and Edwin Gamboa. 2000. Memory of captured prey in three web spiders (Araneae: Araneidae, Linyphiidae, Tetragnathidae). *Animal Cognition* 3: 91-97.

Roland, Chantal, and Jerome S. Rovner. 1983. Chemical and vibratory communication in the aquatic pisaurid spider *Dolomedes triton*. *J. Arachnology* 11: 77-85.

Rovner, Jerome S. 1975. Sound production by nearctic wolf spiders: a substratum-coupled stridulatory mechanism. *Science* 190: 1309-1310.

Rovner, Jerome S., Gaile A. Higashi, and Rainer F. Foelix. 1973. Maternal behavior in wolf spiders: the role of abdominal hairs. *Science* 182: 1153-1155.

Rowell, David M., and Barbara York Main. 1992. Sex ratio in the social spider *Diaea socialis* (Araneae: Thomisidae). *J. Arachnology* 20: 200-206.

Rypstra, Ann L., and R. Scott Tirey. 1989. Observations on the social spider, *Anelosimus domingo* (Araneae, Theridiidae), in southwestern Peru. *J. Arachnology* 17: 368-371.

Schaeffer, Francis A. 1968. *The God Who is There.* Downers Grove, IL: InterVarsity Press.

Schmalhofer, Victoria R. 2000. Diet-induced and morphological color changes in juvenile crab spiders (Araneae, Thomisidae). *J. Arachnology* 28: 56-60.

Schneider, Jutta M. 1999. Delayed oviposition: a female strategy to counter infanticide by males? *Behavioral Ecology* 10(5): 567-571.

Schoener, Thomas W., and David A. Spiller. 1992. Stabilimenta characteristics of the spider *Argiope argentata* on small islands: support of the predator-defense hypothesis. *Behavioral Ecology and Sociobiology* 31: 309-318.

Schuster, Martin, et al. 1994. Field observations on the population structure of three ctenid spiders (*Cupiennius*, Araneae, Ctenidae). *J. Arachnology* 22: 32-38.

Schütz, Dolores, and Michael Taborsky. 2003. Adaptations to an aquatic life may be responsible for the reversed sexual size dimorphism in the water spider, *Argyroneta aquatica*. *Evolutionary Ecology Research* 5: 105-117.

Schwendiger, Peter J. 2003. Two new species of the arboreal trapdoor spider genus *Sason* (Araneae: Barychelidae) from southeast Asia. *The Raffles Bulletin of Zoology* 51(2): 197-207.

Sedey, Kris A., and Elizabeth M. Jakob. 1998. A description of an unusual dome web occupied by egg-carrying *Holocnemus pluchei* (Araneae, Pholcidae). *J. Arachnology* 26: 385-388.

Selden, Paul A. 1989. Orb-web weaving spiders in the early Cretaceous. *Nature* 340: 711-713.

Selden, Paul A. 1996. Fossil mesothele spiders. *Nature* 379: 498-499.

Selden, Paul A. 2002. First British Mesozoic spider, from Cretaceous amber of the Isle of Wight. *Palaeontology* 45(5): 973-983.

Selden, Paul A., et al. 1999. Fossil araneomorph spiders from the Triassic of South Africa and Virginia. *J. Arachnology* 27: 401-414.

Selden, Paul A., Fabio da Costa Casado, and Marisa Vianna Mesquita. 2002. Funnel-web spiders (Araneae: Dipluridae) from the lower Cretaceous of Brazil. *Boletim do 6º Simpósio sobre o Cretácio do Brasil/2do Simposio sobre el Cretácico de América del Sur* (2002): 89-91.

Selden, Paul A., José A. Corronca, and Mario A. Hünicken. 2005. The true identity of the supposed giant fossil spider *Megarachne*. *Biology Letters* 1: 44-48.

Shao, Z., et al. 1999. Analysis of spider silk in native and supercontracted states using Raman spectroscopy. *Polymer* 40: 2493-2500.

Shear, William A., et al. 1989. A Devonian spinneret: early evidence of spiders and silk use. *Science* 246: 479-481.

Shook, Roland S. 1978. Ecology of the wolf spider, *Lycosa carolinensis* Walckenaer (Araneae, Lycosidae) in a desert community. *J. Arachnology* 6: 53-64.

Siegel, George J., M.D., et al. 1998. *Basic Neurochemistry: Molecular, Cellular, and Medical Aspects.* Sixth edition. Philadelphia, PA: Lippincott, Williams, and Wilkins.

Smith, Christopher, et al. 2003. Creep and low strength of spider dragline subjected to constant loads. *J. Arachnology* 31: 421-424.

Smith, Risa B., and Thomas P. Mommsen. 1984. Pollen feeding in an orb-weaving spider. *Science* 226: 1330-1332.

Somers, A. N. 1902. An ingenious spider. *Popular Science News* 36(April): 98.

Stålhandske, Pia. 2002. Nuptial gifts of male spiders function as sensory traps. *Proc. Royal Society London B* 269: 905-908.

Starks, P. T. 2002. The adaptive significance of stabilimenta in orb-webs: a hierarchical approach. *Annales Zoologici Fennici* 39: 307-315.

Stauffer, Steven L., Scott L. Coguill, and Randolph V. Lewis. 1994. Comparison of physical properties of three silks from *Nephila clavipes* and *Araneus gemmoides*. *J. Arachnology* 22: 5-11.

Stowe, Mark K. 1986. Prey specialization in the Araneidae. in William A. Shear, ed. *Spiders: Webs, Behavior, and Evolution*. Stanford, CA: Stanford University Press.

Stowe, Mark K., James H. Tumlinson, and Robert R. Heath. 1987. Chemical mimicry: bolas spiders emit components of moth prey species sex pheromones. *Science* 236: 964-967.

Suter, R. B. 1985. Intersexual competition for food in the bowl and doily spider, *Frontinella pyramitela* (Araneae, Linyphiidae). *J. Arachnology* 13: 61-70.

Suter, Robert B. 1992. Ballooning: data from spiders in freefall indicate the importance of posture. *J. Arachnology* 20: 107-113.

Suter, Robert B. 1999. An aerial lottery: the physics of ballooning in a chaotic atmosphere. *J. Arachnology* 27: 281-293.

Suter, Robert B. 1999. Cheap transport for fishing spiders (Araneae: Pisauridae): the physics of sailing on the water surface. *J. Arachnology* 27: 489-496.

Suter, Robert B., and Jessica Gruenwald. 2000. Spider size and locomotion on the water surface (Araneae: Pisauridae). *J. Arachnology* 28: 300-308.

Suter, Robert B., and Gregg Renkes. 1984. The courtship of *Frontinella pyramitela* (Araneae, Linyphiidae): patterns, vibrations and function. *J. Arachnology* 12: 37-54.

Suter, Robert B., Gail E. Stratton, and Patricia R. Miller. 2004. Taxonomic variation among spiders in the ability to repel water: surface adhesion and hair density. *J. Arachnology* 32(1): 11-21.

Suter, Robert B., et al. 1997. Locomotion on the water surface: propulsive mechanisms of the fisher spider *Dolomedes triton*. *J. Exp. Biology* 200: 2523-2538.

Taylor, Brian B., and Peck, W. B. 1975. A comparison of northern and southern forms of *Phidippus audax* (Hentz) (Araneida, Salticidae). *J. Arachnology* 2: 89-99.

Taylor, Robin M. 2004. Plant nectar contributes to the survival, activity, growth, and fecundity of the nectar-feeding wandering spider *Cheiracanthium inclusum* (Hentz) (Araneae: Miturgidae). Dissertation (Ph. D.). Ohio State University.

Théry, Marc, and Jérôme Casas. 2002. Predator and prey views of spider camouflage. *Nature* 415: 133.

Townley, Mark A., and Edward K. Tillinghast. 2003. On the use of ampullate gland silks by wolf spiders (Araneae, Lycosidae) for attaching the egg sac to the spinnerets and a proposal for defining nubbins and tartipores. *J. Arachnology* 31(2): 209-245.

Tso, I-Min, Chih-Wei Lin, and En-Cheng Yang. 2004. Colourful orb-weaving spiders, *Nephila pilipes*, through a bee's eyes. *J. Experimental Biology* 207: 2631-2637.

Turner, J. Scott, Johannes R. Henschel, and Yael D. Lubin. 1993. Thermal constraints on prey-capture behavior of a burrowing spider in a hot environment. *Behavioral Ecology and Sociobiology* 33(1): 35-43.

Uetz, George W., Jay Boyle, Craig S. Hieber, and R. Stimson Wilcox. 2002. Antipredator benefits of group living in colonial web-building spiders: the 'early warning' effect. *Animal Behaviour* 63: 445-452.

Vetter, Richard S., and Diane K. Barger. 2002. An infestation of 2,055 brown recluse spiders (Araneae: Sicariidae) and no envenomations in a Kansas home: implications for bite diagnoses in nonendemic areas. *Journal of Medical Entomology* 39(6): 948-951.

Vetter, Richard S., and Sean P. Bush. 2002. Reports of presumptive brown recluse spider bites reinforce improbable diagnosis in regions of North America where the spider is not endemic. *Clinical Infectious Diseases* 35(4): 442-445.

Vetter, Richard S., et al. 2003. Distribution of the medically-implicated hobo spider (Araneae: Agelenidae) and a benign congener, *Tegenaria duellica*, in the United States and Canada. *Journal of Medical Entomology* 40(2): 159-164.

Vogelei, A., and R. Greissl. 1989. Survival strategies of the crab spider *Thomisus onustus* Walckenaer 1806 (Chelicerata, Arachnida, Thomisidae). *Oecologia* 80: 513-515.

Vollrath, Fritz, and David P. Knight. 2001. Liquid crystalline spinning of spider silk. *Nature* 410: 541-548.

Watson, Paul J. 1986. Transmission of a female sex pheromone thwarted by males in the spider *Linyphia litigiosa* (Linyphiidae). *Science* 233: 219-221.

Whitcomb, W. H., and R. Eason. 1965. The mating behavior of *Peucetia viridans* (Araneida: Oxyopidae). *The Florida Entomologist* 48(3): 163-168.

Whitehouse, Mary E. A. 1988. Factors influencing specificity and choice of host in *Argyrodes antipodiana* (Theridiidae, Araneae). *J. Arachnology* 16: 349-355.

Whitehouse, Mary E. A. 1991. To mate or fight? Male-male competition and alternative mating strategies in *Argyrodes antipodiana* (Theridiidae, Araneae). *Behavioural Processes* 23: 163-172.

Whitehouse, Mary, et al. 2002. *Argyrodes*: phylogeny, sociality and interspecific interactions—a report on the *Argyrodes* symposium, Badplaas 2001. *J. Arachnology* 30: 238-245.

Wilcox, Stim, and Robert Jackson. 2002. Jumping spider tricksters: deceit, predation, and cognition. in Mark Bokoff, et al., eds. *The Cognitive Animal: Empirical and Theoretical Perspectives on Animal Cognition*. Cambridge, MA: Bradford Books.

Wills, Matthew A. 2001. How good is the fossil record of arthropods? An assessment using the stratigraphic congruence of cladograms. *Geological Journal* 36(3-4): 187-210.

Witt, Peter N., ed. 1982. *Spider Communication: Mechanisms and Ecological Significance*. Princeton, NJ: Princeton University Press.

Wood, Todd Charles, and Megan J. Murray. 2003. *Understanding the Pattern of Life: Origins and Organization of the Species*. Nashville, TN: Broadman & Holman.

Work, Robert W. 1981. A comparative study of the supercontraction of major ampullate silk fibers of orb-web-building spiders (Araneae). *J. Arachnology* 9: 299-308.

Wunderlich, J. 2002. Ant mimicry by spiders and spider-mite interactions preserved in Baltic amber (Arachnida: Acari, Araneae). in *European Arachnology* 2000. S. Toft and N. Scharff, eds. Aarhus, Denmark: Aarhus University Press.

Yáñez, Martha, and Arturo Locht. 1999. Courtship and mating behavior of *Brachypelma klaasi* (Araneae, Theraphosidae). *J. Arachnology* 27: 165-170.

Yeargan, K. V., and L. W. Quate. 1996. Juvenile bolas spiders attract psychodid flies. *Oecologia* 106: 266-271.

Yeargan, K. V., and L. W. Quate. 1997. Adult male bolas spiders retain juvenile hunting tactics. *Oecologia* 112: 572-576.

Young, Orrey P., and Timothy C. Lockley. 1988. Dragonfly predation upon *Phidippus audax* (Araneae: Salticidae). *J. Arachnology* 16: 121-122.

Zimmerman, Manfred, and John R. Spence. 1998. Phenology and life-cycle regulation of the fishing spider *Dolomedes triton* Walckenaer (Araneae, Pisauridae) in central Alberta. *Can. J. Zoology* 76: 295-309.

Zolnerowich, G., and N. V. Horner. 1985. Gnaphosid spiders of north-central Texas (Araneae, Gnaphosidae). *J. Arachnology* 13: 79-85.

Zschokke, Samuel. 1999. Nomenclature of the orb-web. *J. Arachnology* 27: 542-546.

Zschokke, Samuel. 2003. Spider-web silk from the early Cretaceous. *Nature* 424: 636-637.

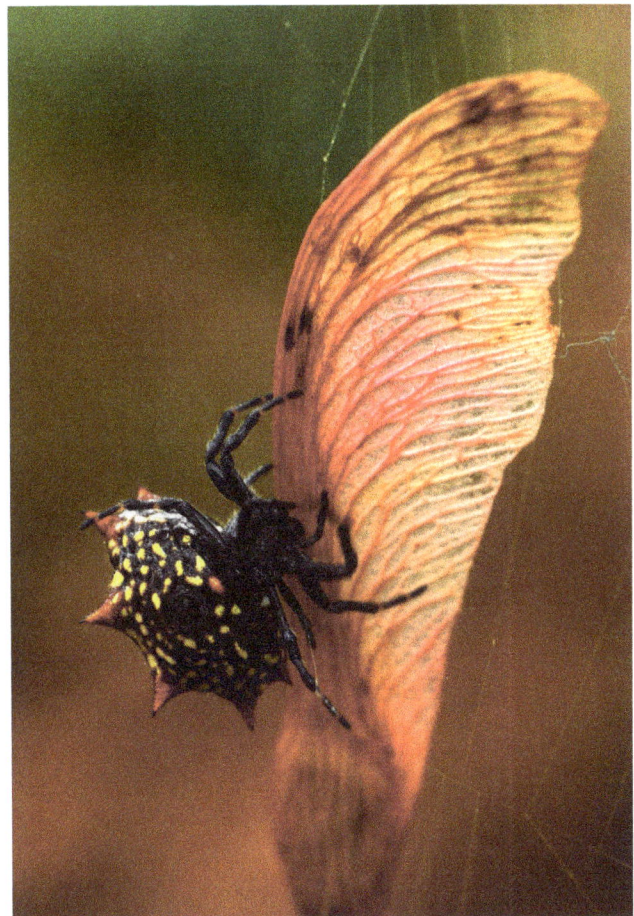

Gasteracantha sp.
CC: Matthew Romack

Beauty and Design in the World of Spiders

www.ingramcontent.com/pod-product-compliance
Lightning Source LLC
Chambersburg PA
CBHW061236270326
41930CB00023B/3491